T0215779

Aspects of Differential Geometry
I

Synthesis Lectures on Mathematics and Statistics

Editor
Steven G. Krantz, *Washington University, St. Louis*

Lectures on Financial Mathematics: Discrete Asset Pricing
Greg Anderson and Alec N. Kercheval
2010

Jordan Canonical Form: Theory and Practice
Steven H. Weintraub
2009

The Geometry of Walker Manifolds
Miguel Brozos-Vázquez, Eduardo García-Río, Peter Gilkey, Stana Nikčević, and Ramón Vázquez-Lorenzo
2009

An Introduction to Multivariable Mathematics
Leon Simon
2008

Jordan Canonical Form: Application to Differential Equations
Steven H. Weintraub
2008

Statistics is Easy!
Dennis Shasha and Manda Wilson
2008

A Gyrovector Space Approach to Hyperbolic Geometry
Abraham Albert Ungar
2008

Aspects of Differential Geometry I

Peter Gilkey, JeongHyeong Park, and Ramón Vázquez-Lorenzo

ISBN: 978-3-031-01279-2   paperback
ISBN: 978-3-031-02407-8   ebook

DOI 10.1007/978-3-031-02407-8

A Publication in the Springer series
SYNTHESIS LECTURES ON MATHEMATICS AND STATISTICS

Lecture #15
Series Editor: Steven G. Krantz, *Washington University, St. Louis*
Series ISSN
Print 1938-1743   Electronic 1938-1751

Aspects of Differential Geometry
I

Peter Gilkey
University of Oregon, Eugene, OR

JeongHyeong Park
Sungkyunkwan University, Suwon, Korea
Institute for Advanced Study, Seoul, Korea

Ramón Vázquez-Lorenzo
University of Santiago de Compostela, Santiago de Compostela, Spain

SYNTHESIS LECTURES ON MATHEMATICS AND STATISTICS #15

ABSTRACT

Differential Geometry is a wide field. We have chosen to concentrate upon certain aspects that are appropriate for an introduction to the subject; we have not attempted an encyclopedic treatment.

In Book I, we focus on preliminaries. Chapter 1 provides an introduction to multivariable calculus and treats the Inverse Function Theorem, Implicit Function Theorem, the theory of the Riemann Integral, and the Change of Variable Theorem. Chapter 2 treats smooth manifolds, the tangent and cotangent bundles, and Stokes' Theorem. Chapter 3 is an introduction to Riemannian geometry. The Levi–Civita connection is presented, geodesics introduced, the Jacobi operator is discussed, and the Gauss–Bonnet Theorem is proved. The material is appropriate for an undergraduate course in the subject.

We have given some different proofs than those that are classically given and there is some new material in these volumes. For example, the treatment of the Chern–Gauss–Bonnet Theorem for pseudo-Riemannian manifolds with boundary is new.

KEYWORDS

Change of Variable Theorem, derivative as best linear approximation, Fubini's Theorem, Gauss–Bonnet Theorem, Gauss's Theorem, geodesic, Green's Theorem, Implicit Function Theorem, improper integrals, Inverse Function Theorem, Levi–Civita connection, partitions of unity, pseudo-Riemannian geometry, Riemann integral, Riemannian geometry, Stokes' Theorem

*This book is dedicated to
Alison, Arnie, Carmen, Junmin,
Junpyo, Manuel, Montse, Rosalía, and Susana.*

Contents

Preface

This two-volume series arose out of work by the three authors over a number of years both in teaching various courses and also in their research endeavors.

The present volume (Book I) is comprised of three chapters. Chapter 1 provides an introduction to multivariable calculus. It begins with two introductory sections on metric spaces and linear algebra. Various notions of differentiability are introduced and the chain rule is proved. The Inverse and Implicit Function Theorems are established. One then turns to the theory of integration. The Riemann integral is introduced and it is shown that a bounded function is integrable if and only if it is integrable almost everywhere. Compact exhaustions by Jordan measurable sets, mesa functions, and partitions of unity are used to define improper integrals. Chapter 1 concludes with a proof of the Change of Variable Theorem; upper and lower sums defined by cubes (rather than rectangles) together with partitions of unity are the fundamental tools employed.

Chapter 2 completes the discussion of multivariable calculus. The basic materials concerning smooth manifolds are introduced. It is shown any compact manifold embeds smoothly in $\mathbb{R}^m$ for some m. A brief introduction to fiber bundle theory and vector bundle theory is given and the tangent and cotangent bundles are introduced. This formalism is then combined with the results of Chapter 1 to establish the generalized Stokes' Theorem. The classical Green's Theorem, Gauss's Theorem, and Stokes' Theorem are then established. The Brauer Fixed Point Theorem, the Fundamental Theorem of Algebra, and the Combing the Hair on a Billiard Ball Theorem are presented as applications.

Chapter 3 presents an introduction to Riemannian and pseudo-Riemannian geometry. The volume form is introduced. The notion of a connection on an arbitrary vector bundle is presented and the discussion is then specialized to the Levi–Civita connection. Geodesics are treated and the classical Hopf–Rinow Theorem giving various equivalent notions of completeness is established in the Riemannian setting. The Jacobi operator is introduced and used to establish the Myers Theorem that if the Ricci tensor on a complete Riemannian manifold is uniformly positive, then the manifold is compact and has finite fundamental group. Riemann surfaces are introduced and the classical Gauss–Bonnet Theorem is established. The Chern–Gauss–Bonnet Theorem in higher dimensions is treated and analytic continuation used to establish an analogous result in the pseudo-Riemannian setting.

We have tried whenever possible to give the original references to major theorems in this area. We have provided a number of pictures to illustrate the discussion, especially in Chapters 1 and 2. Chapters 1 and 2 are suitable for an undergraduate course on "Calculus on Manifolds" and arose in that context out of a course at the University of Oregon. Chapter 3 is designed for an undergraduate course in Differential Geometry. Thus Book I is suitable as an undergraduate text although, of course, it also forms the foundation of a graduate course in Differential Geometry as well. Book II can be used as a graduate text in Differential Geometry and arose in that context out of a second-year graduate course in Differential Geometry at the University of Oregon. The material can, however, also form the basis of a second-semester course at the undergraduate level as well. While much of the material is, of course, standard, many of the proofs are a bit different from those given classically and we hope provide a new viewpoint on the subject. There are also new results in the book; our treatment of the generalized Chern–Gauss–Bonnet Theorem in the indefinite signature context arose out of our study of Euler–Lagrange equations using perturbations of complex metrics (i.e., metrics where the g_{ij} tensor is $\mathbb{C}$-valued). Similarly, our treatment of curves in $\mathbb{R}^m$ given by the solution to constant coefficient ODEs which have finite total curvature is new. There are other examples; Differential Geometry is of necessity a vibrant and growing field – it is not static! There are, of course, many topics that we have not covered – this is a work on "Aspects of Differential Geometry" and of necessity must omit more topics than can possibly be included.

For technical reasons, the material is divided into two books and each book is largely self-sufficient. To facilitate cross references between the two books, we have numbered the chapters of Book I from 1 to 3, and the chapters of Book II from 4 to 8.

Peter Gilkey, JeongHyeong Park, and Ramón Vázquez-Lorenzo
February 2015

Acknowledgments

We have provided many images of famous mathematicians in these two books; mathematics is created by real people and we think having such images makes this point more explicit. The older pictures are in the public domain. We are grateful to the Archives of the Mathematisches Forschungsinstitut Oberwolfach for permitting us to use many images from their archives (R. Brauer, H. Cartan, S. Chern, G. de Rham, S. Eilenberg, H. Hopf, E. Kähler, H. Künneth, L. Nirenberg, H. Poincaré, W. Rinow, L. Vietoris, and H. Weyl); the use of these images was granted to us for the publication of these books only and their further reproduction is prohibited without their express permission. Some of the images (E. Beltrami, E. Cartan, G. Frobenius, and F. Klein) provided to us by the MFO are from the collection of the Mathematische Gesellschaft Hamburg; again, the use of any of these images was granted to us for the publication of these books only and their further reproduction is prohibited without their express permission.

The research of the authors was supported by the Basic Science Research Program through the National Research Foundation of Korea (NRF) funded by the Ministry of Education (2014053413) and by Project MTM2013-41335-P with FEDER funds (Spain). The authors are very grateful to Esteban Calviño-Louzao and Eduardo García-Río for constructive suggestions and assistance in proofreading. It is a pleasure to thank Mr. David Swanson for constructive suggestions concerning the exposition of Chapter 1. The assistance of Ekaterina Puffini of the Krill Institute of Technology has been invaluable. Wikipedia has been a useful guide to tracking down the original references and was a source of many of the older images that we have used that are in the public domain.

Peter Gilkey, JeongHyeong Park, and Ramón Vázquez-Lorenzo
February 2015

CHAPTER 1

Basic Notions and Concepts

In Chapter 1 we present some material from multivariable calculus that will form the foundation of our later discussion of manifold theory. We begin in Section 1.1 by reviewing the theory of metric spaces. Section 1.2 presents some preliminary material concerning linear algebra. Normed vector spaces and inner product spaces are treated as is the norm of a linear map between normed linear spaces. The derivative as the best linear approximation is discussed in Section 1.3 and the chain rule is established. The Inverse and Implicit Function Theorems are proved in Section 1.4. A brief sketch of the theory of Riemann integration is given in Section 1.5. Sets of measure zero and content zero are discussed and Fubini's Theorem is proved. In Section 1.6, improper integrals are introduced using mesa functions and partitions of unity. We conclude Chapter 1 by proving the Change of Variable Theorem in Section 1.7. The calculus was founded by Gottfried Wilhelm von Leibniz and Sir Isaac Newton.

Gottfried Wilhelm von Leibniz (1646–1716)   Sir Isaac Newton (1643–1727)

1.1   METRIC SPACES

We present some basic material concerning metric spaces to establish notation; a good reference would be Baum [6] for point set topology and Rudin [36] for metric spaces. Let X be a set and let $\mathfrak{O}$ be a collection of subsets of X. We say that $\mathfrak{O}$ is a *topology* on X if the following axioms are satisfied:

1. If $\{\mathcal{O}_i\}_{i=1}^n$ is a finite collection of elements of $\mathfrak{O}$ then $\mathcal{O}_1 \cap \cdots \cap \mathcal{O}_n \in \mathfrak{O}$.

2. If $\{\mathcal{O}_\alpha\}_{\alpha \in A}$ is an arbitrary collection of elements of $\mathfrak{O}$, then $\cup_{\alpha \in A}\mathcal{O}_\alpha \in \mathfrak{O}$.

3. The empty set and all of X belong to $\mathfrak{O}$.

A set $\mathcal{O}$ is said to be *open* if and only if it is a member of $\mathfrak{O}$; $\mathfrak{O}$ is the collection of open sets. We say a map f from a topological space X to a topological space Y is *continuous* if $\mathcal{O} \in \mathfrak{O}_Y$ implies $f^{-1}(\mathcal{O}) \in \mathfrak{O}_X$. A set C is said to be *closed* if and only if the complement C^c is open. Consequently, we have equivalently that f is continuous on all of X if and only if $f^{-1}(C)$ is a closed subset of X for any closed set $C \subset Y$. A bijective map f from X to Y is said to be a *homeomorphism* if f and f^{-1} are continuous. A pair (X, d) is said to be a *metric space* and d is said to be the *distance function* if d is a non-negative function on $X \times X$ which satisfies:

1. $d(x, y) = d(y, x) \geq 0$ for all $x, y \in X$.
2. $d(x, y) = 0$ if and only if $x = y$.
3. $d(x, z) \leq d(x, y) + d(y, z)$ for all $x, y, z \in X$.

Let $B_r(x) := \{y \in X : d(x, y) < r\}$ be the *open ball* of radius r about a point x of X. A subset $\mathcal{O}$ of X is said to be *open* if given any point $x \in \mathcal{O}$, there exists $\delta = \delta(x, \mathcal{O}) > 0$ so $B_\delta(x) \subset \mathcal{O}$; this defines a topology on X. A map f from a metric space (X, d_X) to a metric space (Y, d_Y) is *continuous* at $x \in X$ if given $\epsilon > 0$ there exists $\delta = \delta(f, x, \epsilon) > 0$ so $f(B_\delta(x)) \subset B_\epsilon(f(x))$; f is continuous on all of X if it is continuous at every point of X. If S is a subset of a metric space X, then the restriction of the distance on X to S makes S a metric space; the inclusion map of S in X is then a continuous map. The Euclidean distance function on $\mathbb{R}^m$ is given by:

$$d((x^1, \ldots, x^m), (y^1, \ldots, y^m)) := \{(x^1 - y^1)^2 + \cdots + (x^m - y^m)^2\}^{\frac{1}{2}}.$$

Let (X, d) be a metric space. Given a sequence of points $x_n \in X$, we say $x_n \to x$ as $n \to \infty$ and write $\lim_{n \to \infty} x_n = x$ if given $\epsilon > 0$, there exists $N = N(\epsilon)$ so $n \geq N$ implies $d(x, x_n) < \epsilon$; x is said to be the *limit point* of the sequence. A point x is said to be a *cluster point* of a set S if there exists a sequence of points $x_n \in S$ so $\lim_{n \to \infty} x_n = x$. Let S' be the set of cluster points of S. The set S is said to be *closed* if $S' \subset S$. We let $\bar{S} = S \cup S'$ be the *closure* of S; $\bar{S}$ is a closed set since $(S \cup S')' = S'$. For example, the closure of the open ball is the *closed ball*

$$\bar{B}_r(x) := \{y \in X : d(x, y) \leq r\}.$$

We say that a set C is *bounded* if there exists (r, x) so $C \subset B_r(x)$. We say that a subset C of X is *compact* if given any open cover $\{\mathcal{O}_\alpha\}$ of C, there exists a finite subcover. The following is well-known:

Lemma 1.1   Let X and Y be metric spaces.

1. If C is a compact subset of X, then C is closed and bounded.

2. If C is a closed subset of a compact set, then C is compact.

3. If C is a closed and bounded subset of $\mathbb{R}^m$, then C is compact.

4. If $\{x_n\}$ is a sequence of points in a compact set, then there is a convergent subsequence.

5. A continuous real-valued function on a compact set attains its maximum and minimum values.

6. If X is compact and if f is a continuous map from X onto Y, then Y is compact.

7. If f is a continuous bijective map from X to Y and if X is compact, then f is a homeomorphism.

1.2   LINEAR ALGEBRA

1.2.1 THE REAL, COMPLEX, AND QUATERNION FIELDS. We shall let $\mathbb{R}$, $\mathbb{C}$ and $\mathbb{H} := \{x_0 + ix_1 + jx_2 + kx_3\}$ be the real, complex, and the quaternion numbers, respectively, where we have $i^2 = j^2 = k^2 = -1$, $ij = -ji = k$, $jk = -kj = i$, and $ki = -ik = j$. Both $\mathbb{C}$ and $\mathbb{H}$ have a *conjugation operator*. We set

$$\overline{x + iy} = x - iy \quad \text{and} \quad \overline{x_0 + ix_1 + jx_2 + kx_3} = x_0 - ix_1 - jx_2 - kx_3.$$

Let $\|\cdot\|$ be the usual Euclidean norm. Let $z \in \mathbb{C} = \mathbb{R}^2$ and let $w \in \mathbb{H} = \mathbb{R}^4$. Then:

$$\|z\|^2 = z\bar{z} = \bar{z}z = x^2 + y^2 \qquad \text{so} \quad z^{-1} = \|z\|^{-2}\bar{z},$$
$$\|w\|^2 = w\bar{w} = \bar{w}w = x_0^2 + x_1^2 + x_2^2 + x_3^2 \quad \text{so} \quad w^{-1} = \|w\|^{-2}\bar{w}.$$

If $z_i \in \mathbb{C}$ and $w_i \in \mathbb{H}$, then $\overline{z_1 z_2} = \bar{z}_1\, \bar{z}_2$ and $\overline{w_1 w_2} = \bar{w}_2\, \bar{w}_1$.

1.2.2 VECTOR SPACES. Let V be a finite-dimensional $\mathbb{F}$ vector space of dimension m; if we omit the field, it is assumed to be $\mathbb{R}$. For example, in what follows $\mathrm{GL}(V) = \mathrm{GL}_{\mathbb{R}}(V)$. The dual bundle V^* is the vector space of linear maps from V to $\mathbb{F}$. If $\{e_1, \ldots, e_m\}$ is a basis for V, we can expand any element $x \in V$ in the form $x^i e_i = x^1 e_1 + \cdots + x^m e_m$ where we adopt the *Einstein convention* and sum over repeated indices. Let $e^i(x) := x^i$. The $\{e^i\}$ form a basis for V^* that is called the *dual basis* and the $\{x^i\}$ are called the *coordinate functions*. The map $x \to (x^1, \ldots, x^m)$ defines a vector space isomorphism from V to $\mathbb{F}^m$.

1.2.3 NORMED VECTOR SPACES. A pair $(V, \|\cdot\|)$ is said to be a *normed vector space* if V is a real vector space and if $\|\cdot\|$ is a map from V to $\mathbb{R}$ with:

$$\|x\| \geq 0 \text{ and } \|x\| = 0 \text{ if and only if } x = 0 \text{ for all } x \in V, \tag{1.2.a}$$
$$\|cx\| = |c|\,\|x\| \text{ for all } x \in V \text{ and for all } c \in \mathbb{R}, \tag{1.2.b}$$
$$\|x + y\| \leq \|x\| + \|y\| \text{ for all } x, y \in V. \tag{1.2.c}$$

We may give V the structure of a metric space by setting $d(x, y) = \|x - y\|$ and thereby define a topology on V. Let $\langle \cdot, \cdot \rangle$ be a positive definite symmetric bilinear form on V; the pair $(V, \langle \cdot, \cdot \rangle)$ is said to be a *Euclidean vector space* and we set:

$$\|x\| := \langle x, x \rangle^{\frac{1}{2}}.$$

The properties of Equation (1.2.a) and Equation (1.2.b) are then immediate. One has the *Cauchy–Schwarz–Bunyakovsky inequality*

$$|\langle x, y \rangle| \le \|x\| \cdot \|y\| \quad \text{for all} \quad x, y \in V .$$

V. Bunyakovsky (1804–1889)   A. Cauchy (1789–1857)   K. Schwarz (1843–1921)

The triangle inequality of Equation (1.2.c) is immediate from this identity; this completes the verification that $(V, \|\cdot\|)$ is a normed linear space in this setting.

Let $(V, \langle \cdot, \cdot \rangle)$ be a Euclidean vector space of dimension m over $\mathbb{R}$. We can use the *Gram–Schmidt process* to find a basis $\{e_1, \dots, e_m\}$ for V so that $\langle e_i, e_j \rangle = \delta_{ij}$ where δ is the *Kronecker symbol*

$$\delta_{ij} := \left\{ \begin{array}{ll} 0 & \text{if} \quad i \ne j \\ 1 & \text{if} \quad i = j \end{array} \right\} .$$

Such a basis is said to be an *orthonormal basis*. Let $x = x^i e_i$. Then:

$$x^i = \langle x, e_i \rangle, \quad \langle x, y \rangle = \sum_i x^i y^i \quad \text{and} \quad \|x\| = \left\{ \sum_i (x^i)^2 \right\}^{\frac{1}{2}} .$$

Thus the map $T : x \to (x^1, \dots, x^m)$ is a linear isomorphism from V to $\mathbb{R}^m$ which preserves the inner products, i.e., $\langle x, y \rangle_V = \langle Tx, Ty \rangle_{\mathbb{R}^m}$; such a map is called an *isometry*. Note that:

$$|x^i| \le \|x\| \quad \text{for} \quad 1 \le i \le m . \tag{1.2.d}$$

We have the *polarization identity* which expresses the inner product in terms of the norm:

$$\langle x, y \rangle = \tfrac{1}{4}\{\|x + y\|^2 - \|x - y\|^2\} .$$

We say that two norms $\|\cdot\|_1$ and $\|\cdot\|_2$ on V are *equivalent norms* if there exist positive constants C_1 and C_2 so that $\|x\|_1 \le C_1 \|x\|_2$ and $\|x\|_2 \le C_2 \|x\|_1$ for all $x \in V$; equivalent norms define the same topology. The following is a useful observation:

Lemma 1.2   Let V be a finite-dimensional real vector space.

1. Any two norms on V are equivalent.

2. Let $\|\cdot\|_1$ be a norm on V. Let $\mathcal{B} = \{e_1, \dots, e_m\}$ be a basis for V. There exists a constant $C = C(\|\cdot\|_1, \mathcal{B})$ so that if we expand $x = x^1 e_1 + \cdots + x^m e_m$, then $|x^i| \leq C\|x\|_1$.

Proof. Let $\|\cdot\|$ be an auxiliary norm on V which is given by a positive definite inner product $\langle \cdot, \cdot \rangle$ on V. We will show $\|\cdot\|_1$ is equivalent to $\|\cdot\|$; a similar argument shows $\|\cdot\|_2$ is equivalent to $\|\cdot\|$ and hence $\|\cdot\|_1$ is equivalent to $\|\cdot\|_2$. Let $\{e_1, \dots, e_m\}$ be an orthonormal basis for V with respect to $\langle \cdot, \cdot \rangle$; we identify V with $\mathbb{R}^m$. We may use Equation (1.2.d) to estimate:

$$
\begin{aligned}
\|x - y\|_1 &= \|(x^1 - y^1)e_1 + \cdots + (x^m - y^m)e_m\|_1 \\
&\leq |x^1 - y^1|\,\|e_1\|_1 + \cdots + |x^m - y^m|\,\|e_m\|_1 \\
&\leq \|x - y\|\{\|e_1\|_1 + \cdots + \|e_m\|_1\} \leq \kappa\|x - y\|
\end{aligned}
$$

for $\kappa := \|e_1\|_1 + \cdots + \|e_m\|_1$. Given $\epsilon > 0$, let $\delta = \epsilon/(\kappa + 1)$. If $d(x, y) = \|x - y\| < \delta$, then $\|x - y\|_1 < \delta\kappa < \epsilon$. By the triangle inequality,

$$
\|x\|_1 \leq \|y\|_1 + \|y - x\|_1 \quad \text{and} \quad \|y\|_1 \leq \|x\|_1 + \|x - y\|_1 \,.
$$

Consequently, $\big| \|x\|_1 - \|y\|_1 \big| \leq \|x - y\|_1 < \epsilon$. This shows that the map $x \rightarrow \|x\|_1$ is continuous with respect to the topology defined by $\|\cdot\|$ (which is the usual Euclidean topology on $\mathbb{R}^m$). We apply Lemma 1.1. The unit sphere in $\mathbb{R}^m$ is compact. A continuous function on a compact set attains its minimum and maximum values. Let

$$
\tilde{C}_1 := \min_{\|x\|=1} \|x\|_1 \quad \text{and} \quad \tilde{C}_2 := \max_{\|x\|=1} \|x\|_1 \,.
$$

Since $\|x\|_1 = 0$ if and only if $x = 0$, $\tilde{C}_1$ and $\tilde{C}_2$ are positive. The desired inequalities of Assertion 1 for all x now follow from the following identity by rescaling:

$$
\tilde{C}_1\|x\| \leq \|x\|_1 \leq \tilde{C}_2\|x\| \quad \text{for all} \quad x \in V \quad \text{with} \quad \|x\| = 1 \,.
$$

Let $\mathcal{B}$ be a basis for V. We may define a positive definite inner product on V by requiring the basis to be orthonormal and obtain an auxiliary inner product $\langle \cdot, \cdot \rangle$ with associated norm $\|\cdot\|_2$. We use Equation (1.2.d) to estimate $|x^i| \leq \|x\|_2$. The desired estimate using $\|\cdot\|_1$ now follows from Assertion 1 as $\|\cdot\|_1$ is equivalent to $\|\cdot\|_2$. $\qquad\square$

If $(V, \|\cdot\|)$ is a finite-dimensional real normed linear space, let $d_{\|\cdot\|}(x, y) := \|x - y\|$. By Lemma 1.2, any two norms are equivalent. Consequently, the underlying topology on V is independent of the particular norm which is chosen; the identity map is a homeomorphism from $(V, d_{\|\cdot\|_1})$ to $(V, d_{\|\cdot\|_2})$ for any pair of norms on V. Furthermore, the coordinate functions x^i relative to a basis for V are continuous with respect to this topology.

Let $(V, \|\cdot\|_V)$ and $(W, \|\cdot\|_W)$ be finite-dimensional normed vector spaces. Let A be a linear map from V to W, and let $\{e_1, \ldots, e_m\}$ be a basis for V. By Lemma 1.2:

$$
\begin{aligned}
\|Ax\|_W &= \|x^1 Ae_1 + \cdots + x^m Ae_m\|_W \\
&\leq |x^1| \|Ae_1\|_W + \cdots + |x^m| \|Ae_m\|_W \\
&\leq C\|x\|_V \{\|Ae_1\|_W + \cdots + \|Ae_m\|_W\} \\
&= C\kappa\|x\|_V \quad \text{for} \quad \kappa := \|Ae_1\|_W + \cdots + \|Ae_m\|_W .
\end{aligned}
\tag{1.2.e}
$$

If $\epsilon > 0$ is given, let $\delta = \epsilon/(2C\kappa + 1)$. By Equation (1.2.e), if $\|x - y\|_V < \delta$, then we have that $\|Ax - Ay\|_W = \|A(x - y)\|_W \leq C\kappa\|x - y\|_V < \epsilon$. Consequently, A is a continuous map from V to W. Equation (1.2.e) also shows that

$$
\|A\| := \sup_{0 \neq x} \frac{\|Ax\|_W}{\|x\|_V} \leq C\kappa
\tag{1.2.f}
$$

is well-defined and finite. We then have

$$
\|Ax\|_W \leq \|x\|_V \|A\| \quad \text{for all} \quad x \in V .
$$

Let $\mathrm{Hom}(V, W)$ be the set of all linear maps from V to W; $(\mathrm{Hom}(V, W), \|\cdot\|)$ is a normed vector space with

$$
\dim(\mathrm{Hom}(V, W)) = \dim(V) \dim(W) .
$$

If we choose a basis, we may identify V with $\mathbb{R}^m$ and $\mathrm{End}(V, V)$ with the set of $m \times m$ matrices $M_m(\mathbb{R})$. If $A \in M_m(\mathbb{R})$, let $\det(A)$ be the determinant of A.

Lemma 1.3

1. **(Cramer's Rule [13])** Let $A \in \mathrm{Hom}(\mathbb{R}^m, \mathbb{R}^m)$ be an $m \times m$ matrix.

    (a) Then A is invertible if and only if $\det(A) \neq 0$.

    (b) Let C_{ij} be the matrix of cofactors formed by crossing out row i and column j. If $\det(A) \neq 0$, then $(A^{-1})_{ij} = (-1)^{i+j} C_{ji} \det(A)^{-1}$.

2. Let $A \in \mathrm{Hom}(V, W)$. If $\lim_{x \to 0} \frac{\|Ax\|_W}{\|x\|_V} = 0$, then $A = 0$.

Proof. Assertion 1 is well-known so we will omit the proof. Suppose $A \neq 0$. Choose $v \in V$ so $Av \neq 0$. Let $v_n := v/n$. Then $v_n \to 0$ so

$$
0 = \lim_{n \to \infty} \frac{\|Av_n\|_W}{\|v_n\|_V} = \lim_{n \to \infty} \frac{\frac{1}{n}\|Av\|_W}{\frac{1}{n}\|v\|_V} = \frac{\|Av\|_W}{\|v\|_V} \neq 0 .
$$

This contradiction establishes the lemma. $\qquad \square$

1.2.4 THE GENERAL LINEAR GROUP. If $\mathbb{F} \in \{\mathbb{R}, \mathbb{C}, \mathbb{H}\}$, then the *general linear group* $\mathrm{GL}_{\mathbb{F}}(V)$ is the set of all invertible $\mathbb{F}$ linear transformations from V to V. Suppose that $\mathbb{F} = \mathbb{R}$ or that $\mathbb{F} = \mathbb{C}$. Since the determinant function det is continuous, $\mathrm{GL}_{\mathbb{F}}(V)$ is an open subset of $\mathrm{Hom}_{\mathbb{F}}(V, V)$; $\mathrm{GL}_{\mathbb{F}}(V)$ forms group under composition. We use Cramer's rule to see that the group operation and inverse are continuous with respect to the induced topology. If $\mathbb{F} = \mathbb{H}$, then we cannot use the determinant directly as $\mathbb{H}$ is non-commutative. We can forget the quaternion structure to define an underlying real vector space $V_{\mathbb{R}}$. If $A \in \mathrm{Hom}_{\mathbb{H}}(V, V)$, then A defines an element $A_{\mathbb{R}}$ of $\mathrm{Hom}_{\mathbb{R}}(V_{\mathbb{R}}, V_{\mathbb{R}})$. We have

$$A \text{ is invertible } \Leftrightarrow A \text{ is injective } \Leftrightarrow A_{\mathbb{R}} \text{ is injective } \Leftrightarrow A_{\mathbb{R}} \text{ is invertible } \Leftrightarrow \det(A_{\mathbb{R}}) \neq 0.$$

Thus $\mathrm{GL}_{\mathbb{H}}(V) = \{A \in \mathrm{Hom}_{\mathbb{H}}(V, V) : \det(A_{\mathbb{R}}) \neq 0\}$. This shows $\mathrm{GL}_{\mathbb{H}}(V)$ is an open subset of $\mathrm{Hom}_{\mathbb{H}}(V, V)$. Furthermore, since $A_{\mathbb{R}} \to A_{\mathbb{R}}^{-1}$ is continuous on $\mathrm{GL}_{\mathbb{R}}(V_{\mathbb{R}})$, the map $A \to A^{-1}$ is continuous on the subgroup $\mathrm{GL}_{\mathbb{H}}(V) \subset \mathrm{GL}_{\mathbb{R}}(V_{\mathbb{R}})$.

1.3   THE DERIVATIVE

In this section, we will establish the basic facts concerning the differential calculus that we shall need. A good auxiliary reference would be Spivak [38]. The appropriate extension of the notion of a derivative to multivariable calculus is as the best linear approximation. We begin by presenting several different notions of differentiability:

Definition 1.4   Let $(V, \| \cdot \|)$ and $(W, \| \cdot \|)$ be finite-dimensional real normed vector spaces, with $\dim(V) = m$ and $\dim(W) = n$. Let $\mathcal{O}$ be an open subset of V, let $F : \mathcal{O} \to W$, and let P be a point of $\mathcal{O}$.

1. F is said to be *differentiable* at P if there exists a linear map $A : V \to W$ so that

$$\lim_{u \to 0} \frac{\|F(P + u) - F(P) - Au\|}{\|u\|} = 0$$

   or, in other words, $F(P + u) = F(P) + Au + o(\|u\|)$. We note that if such an A exists, then it is unique by Lemma 1.3, and thus we shall call A the *derivative* of F at P and write $A = F'(P)$. Define $\mathcal{E}_F(P_1, P)$ by the identity:

$$F(P_1) = F(P) + A(P_1 - P) + \mathcal{E}_F(P_1, P).$$

   Then $F'(P) = A$ if and only if for any $\epsilon > 0$, there exists $\delta = \delta(\epsilon)$ so $\|P - P_1\| < \delta$ implies

$$\|\mathcal{E}_F(P_1, P)\| \leq \epsilon \|P_1 - P\|.$$

2. F has *a directional derivative* $(D_u F)(P)$ at P in the direction $u \in V$, if the following limit exists:

$$(D_u F)(P) := \lim_{t \to 0} \frac{F(P + tu) - F(P)}{t}.$$

3. Fix a basis $\{e_1, \ldots, e_m\}$ for V and let $(x^1, \ldots, x^m)$ be the dual coordinates on V. If the directional derivative $(D_{e_i} F)(P)$ exists for $1 \leq i \leq m$, then we shall denote this directional derivative by the *partial derivative* $\frac{\partial F}{\partial x^i}(P)$. We shall also use the notation $(\partial_{x^i} F)(P)$. Choose a basis $\{\tilde{e}_k\}$ for W and expand $F = \sum_k F^k \tilde{e}_k$. If all the partial derivatives of F exist at P, we shall denote the *Jacobian matrix* by

$$\begin{pmatrix} \partial_{x^1} F^1 & \ldots & \partial_{x^m} F^1 \\ \ldots & \ldots & \ldots \\ \partial_{x^1} F^n & \ldots & \partial_{x^m} F^n \end{pmatrix} (P).$$

4. We say that F is C^k if F has continuous partial derivatives up to order k. Note that F is C^k for $k \geq 2$ if and only if F' is C^{k-1} where we regard F' as a map from $\mathcal{O}$ to $\mathrm{Hom}(V, W)$. If f is C^k for all k, then f will be said to be *smooth* or C^∞.

The following result is basic in this subject; we fix a system of coordinates and work in Euclidean space for simplicity.

Theorem 1.5   *Let $\mathcal{O}$ be an open subset of $\mathbb{R}^m$ and let $F : \mathcal{O} \to \mathbb{R}^n$.*

1. *If F has continuous partial derivatives at each point $P \in \mathcal{O}$, then F is differentiable at each point P of $\mathcal{O}$.*

2. *If F is differentiable at some point $P \in \mathcal{O}$, then:*

   (a) F is continuous at P.

   (b) All the partial derivatives of F exist at P and $(D_u F)(P) = F'(P)(u)$.

   (c) $F'(P)$ is given by the Jacobian matrix.

3. *If all the directional derivatives of F exist at some point $P \in \mathcal{O}$, then all the partial derivatives of F exist at P.*

Proof. It is immediate from the definition that $F : \mathbb{R}^m \to \mathbb{R}^n$ is differentiable if and only if each of the component functions is differentiable. Thus we shall take $n = 1$ in the proof of Assertion 1. We shall also take $m = 2$ in the interests of notational simplicity; the general case is no more difficult. We assume $\mathcal{O}$ is an open rectangle as this is a local result. We apply the Mean Value Theorem; it was for this reason that we assumed $n = 1$. Let $P = (a, b)$ and let $P_1 = (c, d)$ be points of $\mathcal{O}$. As $\partial_{x^1} F$ and $\partial_{x^2} F$ are continuous,

$$\begin{aligned} F(P_1) - F(P) &= F(c, d) - F(a, b) = F(c, d) - F(a, d) + F(a, d) - F(a, b) \\ &= \partial_{x^1} F(\alpha, d)(c - a) + \partial_{x^2} F(a, \beta)(d - b) \end{aligned}$$

where $\alpha = \alpha(d)$ is between a and c and where $\beta = \beta(a)$ is between b and d. Let A be the Jacobian; $A(u, v) = \partial_{x^1} F(a, b)u + \partial_{x^2} F(a, b)v$. Then:

$$F(P_1) - F(P) - A(P_1 - P) = \mathcal{E}_F(P_1, P) \quad \text{where}$$
$$\mathcal{E}_F(P_1, P) := \{\partial_{x^1} F(\alpha, d) - \partial_{x^1} F(a, b)\}(c - a)$$
$$+ \{\partial_{x^2} F(a, \beta) - \partial_{x^2} F(a, b)\}(d - b).$$

We estimate:

$$\|\mathcal{E}_F(P_1, P)\| \leq \mathcal{E}_1(P_1, P) \|P_1 - P\| \quad \text{where}$$
$$\mathcal{E}_1(P_1, P) := \{|\partial_{x^1} F(\alpha, d) - \partial_{x^1} F(a, b)| + |\partial_{x^2} F(a, \beta) - \partial_{x^2} F(a, b)|\}.$$

Since the partial derivatives are continuous, $\mathcal{E}_1$ can be made arbitrarily small if P_1 is close to P. Assertion 1 now follows.

Suppose F is differentiable at P. We may express

$$\|F(P_1) - F(P) - F'(P)(P_1 - P)\| \leq \mathcal{E}_1 \|P_1 - P\|$$

where $\mathcal{E}_1 = \mathcal{E}_1(P_1, P)$ can be made arbitrarily small for P_1 close to P. In particular, by shrinking $\mathcal{O}$ if necessary, we may assume $\mathcal{E}_1$ is at most 1. We may then estimate

$$\|F(P_1) - F(P)\| \leq \|F'(P)(P_1 - P)\| + \mathcal{E}_1 \|P_1 - P\| \leq \{\|F'(P)\| + 1\} \|P_1 - P\|.$$

This shows that F is continuous at P. Let $0 \neq u$. Then

$$\lim_{t \to 0} \left\{ \frac{F(P + tu) - F(P)}{t} - F'(P)(u) \right\} = \|u\| \lim_{t \to 0} \frac{F(P + tu) - F(P) - F'(P)(tu)}{t \|u\|} = 0.$$

This shows that $(D_u F)(P) = F'(P)(u)$. Thus F has all directional derivatives and in particular partial derivatives at P. Since $F'(P)$ is linear and $F'(P)(e_i) = (D_{e_i} F)(P) = (\partial_{x^i} F)(P)$, F' is the Jacobian matrix. Assertion 2 now follows. Assertion 3 is immediate. □

1.3.1 EXAMPLES. None of the implications of Theorem 1.5 are reversible. We first give an example of a function which is differentiable but not continuously differentiable when $m = 1$.

$$f(x) = \left\{ \begin{array}{ll} 10x^2 \sin(x^{-1}) & \text{if } x \neq 0 \\ 0 & \text{if } x = 0 \end{array} \right\}. \qquad \text{Picture:}$$

We compute:

$$f'(x) = \left\{ \begin{array}{ll} 20x \sin(x^{-1}) - 10 \cos(x^{-1}) & \text{if } x \neq 0 \\ \lim_{x \to 0} \frac{10x^2 \sin(x^{-1}) - 0}{x} = 0 & \text{if } x = 0 \end{array} \right\}.$$

Thus f is differentiable at every point $x \in \mathbb{R}$. However since $\lim_{x \to 0} f'(x)$ does not exist, f is not continuously differentiable.

Next we give an example of a real-valued function with directional derivatives at each point of $\mathbb{R}^2$ but which is nevertheless not differentiable at 0. Set

$$f(x, y) := \begin{cases} \frac{xy^2}{x^2+y^2} & \text{if } (x, y) \neq (0,0) \\ 0 & \text{if } (x, y) = (0,0) \end{cases} . \qquad \text{Picture:}$$

It is clear that f has continuous partial derivatives for $(x, y) \neq (0,0)$ and thus is continuously differentiable for $(x, y) \neq (0,0)$. It is also clear f is continuous at $(0,0)$. Clearly $(D_0 f)(0,0) = 0$. If $u = (x, y) \neq (0,0)$, then

$$(D_u f)(0,0) = \lim_{t \to 0} \frac{t^3 xy^2}{t^3(x^2 + y^2)} = f(x, y).$$

Thus f has directional derivatives at $(0,0)$. We have

$$(D_{e_1} f)(0,0) = (D_{e_2} f)(0,0) = 0.$$

If f was differentiable at 0, then the map which sends $u \to (D_u f)(0,0)$ would be a linear function of u. This implies $(D_u f)(0,0) = x(D_{e_1} f)(0,0) + y(D_{e_2} f)(0,0) = 0$. Since $D_{(1,1)} f(0,0) = f(1, 1) = \frac{1}{2} \neq 0$, this is false. Thus f is not differentiable at the origin.

We could also consider the function

$$f(x, y) := \begin{cases} \frac{xy^2}{x^2+y^4} & \text{if } (x, y) \neq (0,0) \\ 0 & \text{if } (x, y) = (0,0) \end{cases} . \qquad \text{Picture:}$$

We see that f is not continuous as

$$\lim_{t \to 0} f(t^2, t) = \lim_{t \to 0} \frac{t^4}{t^4 + t^4} = \frac{1}{2} \neq f(0,0).$$

If $x = 0$, then $(D_{(0,y)} f)(0,0) = 0$. If $x \neq 0$, then

$$(D_{(x,y)} f)(0,0) = \lim_{t \to 0} \frac{t^3 xy^2}{t^3 x^2 + t^5 y^4} = \frac{y^2}{x}.$$

Thus all the directional derivatives exist and yet f is not continuous.

Finally, we can give an example of a function all of whose partial derivatives exist but does not have directional derivatives in all directions. Let

$$f(x,y) = \begin{cases} \frac{xy}{x^2+y^2} & \text{if} \quad (x,y) \neq (0,0) \\ 0 & \text{if} \quad (x,y) = (0,0) \end{cases}. \quad \text{Picture:}$$

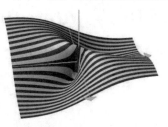

Then all the partial derivatives of f exist for $(x,y) \neq (0,0)$ and it is an easy direct calculation that $\frac{\partial f}{\partial x}(0,0) = \frac{\partial f}{\partial y}(0,0) = 0$. Thus the partial derivatives of f also exist at the origin. However

$$D_{(1,1)}f(0,0) = \lim_{t \to 0} \frac{t^2}{2t^3}$$

does not exist. Thus the partial derivatives can exist without f having directional derivatives at $(0,0)$.

In view of Theorem 1.5, all the usual rules of calculus pertain; the derivative of the sum is the sum of the derivatives, multiplying by a constant rescales the derivative, etc. Furthermore, if $f(x) = Ax$ where A is linear, then f is differentiable and $f'(P) = A$ for any P. There is a delicate point here. If f is an $\mathbb{R}$-valued function of one variable, then $f'(P)$ is usually thought of as a number $\lambda \in \mathbb{R}$. However, in our context we shall think of $f'(P) = m_\lambda$ being the linear transformation defined as multiplication by λ. It is a minor point, but an important one.

1.3.2 THE CHAIN RULE. We now establish the *chain rule*; to simplify the exposition, we assume the functions in question have domains all of $\mathbb{R}^m$ and $\mathbb{R}^n$; this is inessential as all the computations are purely local.

Theorem 1.6   (Chain Rule).   *Let $f : \mathbb{R}^m \to \mathbb{R}^n$ be differentiable at P and let $g : \mathbb{R}^n \to \mathbb{R}^p$ be differentiable at $Q = f(P)$. Then $h := g \circ f : \mathbb{R}^m \to \mathbb{R}^p$ is differentiable at P and the derivative of h at P is given by $h'(P) = g'(Q) \circ f'(P)$.*

Proof. Define $\mathcal{E}_f$, $\mathcal{E}_g$, and $\mathcal{E}_h$ by the equations:

$$f(P_1) = f(P) + f'(P)(P_1 - P) + \mathcal{E}_f(P_1, P),$$
$$g(Q_1) = g(Q) + g'(Q)(Q_1 - Q) + \mathcal{E}_g(Q_1, Q),$$
$$h(P_1) = h(P) + g'(Q)f'(P)(P_1 - P) + \mathcal{E}_h(P_1, P).$$

The assumption that f is differentiable at P and that g is differentiable at $Q = f(P)$ implies that given $\epsilon_f > 0$ and $\epsilon_g > 0$, there exist $\delta_f = \delta_f(\epsilon_f) > 0$ and $\delta_g = \delta_g(\epsilon_g) > 0$ so that

$$\|\mathcal{E}_f(P_1, P)\| \leq \epsilon_f \|P_1 - P\| \quad \text{if} \quad \|P_1 - P\| < \delta_f,$$
$$\|\mathcal{E}_g(Q_1, Q)\| \leq \epsilon_g \|Q_1 - Q\| \quad \text{if} \quad \|Q_1 - Q\| < \delta_g.$$

We have $h(P_1) = g(f(P_1))$ and $h(P) = g(f(P))$. Consequently, we may expand

$$
\begin{aligned}
\mathcal{E}_h(P_1, P) &= g(f(P_1)) - g(f(P)) - g'(Q)f'(P)(P_1 - P) \\
&= g'(Q)(f(P_1) - f(P)) + \mathcal{E}_g(f(P_1), f(P)) - g'(Q)f'(P)(P_1 - P) \\
&= g'(Q)(f'(P)(P_1 - P) + \mathcal{E}_f(P_1, P)) \\
&\quad + \mathcal{E}_g(f(P_1), f(P)) - g'(Q)f'(P)(P_1 - P) \\
&= g'(Q)\mathcal{E}_f(P_1, P) + \mathcal{E}_g(f(P_1), f(P)) \,.
\end{aligned}
$$

Let $\epsilon_h > 0$ be given. Choose $0 < \epsilon_f < 1$ and $0 < \epsilon_g < 1$ so

$$
\|g'(Q)\|\epsilon_f < \tfrac{1}{2}\epsilon_h \quad \text{and} \quad \epsilon_g\{\|f'(P)\| + 1\} < \tfrac{1}{2}\epsilon_h \,.
$$

Let $\delta_f = \delta_f(\epsilon_f)$ and $\delta_g = \delta_g(\epsilon_g)$. Since f is continuous at P, we may choose $0 < \delta < \delta_f$ so

$$
\|P_1 - P\| < \delta \quad \text{implies} \quad \|f(P_1) - f(P)\| < \delta_g \,.
$$

Let $\|P_1 - P\| < \delta$. We estimate

$$
\begin{aligned}
&\|g'(Q)\mathcal{E}_f(P_1, P)\| \leq \|g'(Q)\| \cdot \|\mathcal{E}_f(P_1, P)\| \leq \|g'(Q)\|\epsilon_f\|P_1 - P\| < \tfrac{1}{2}\epsilon_h\|P_1 - P\|, \\
&\|\mathcal{E}_g(f(P_1), f(P))\| \leq \epsilon_g\|f(P_1) - f(P)\| \leq \epsilon_g\{\|f'(P)(P_1 - P)\| + \|\mathcal{E}_f(P_1, P)\|\} \\
&\qquad \leq \epsilon_g\{\|f'(P)\| + \epsilon_f\}\|P_1 - P\| \leq \epsilon_g\{\|f'(P)\| + 1\}\|P_1 - P\| \leq \tfrac{1}{2}\epsilon_h\|P_1 - P\|, \\
&\|\mathcal{E}_h(P_1, P)\| \leq \|g'(Q)\mathcal{E}_f(P_1, P)\| + \|\mathcal{E}_g(f(P_1), f(P))\| \leq \epsilon_h\|P_1 - P\| \,.
\end{aligned}
$$

This shows that h is differentiable at P and that $h'(P) = g'(Q) \circ f'(P)$. $\qquad\square$

1.3.3 INDEX NOTATION. We introduce the standard *index notation*. Let

$$
x = (x^1, \ldots, x^m), \quad y = (y^1, \ldots, y^n), \quad \text{and} \quad z = (z^1, \ldots, z^p)
$$

be coordinates on $\mathbb{R}^m$, on $\mathbb{R}^n$, and on $\mathbb{R}^p$, respectively. If A is a linear map from $\mathbb{R}^p$ to $\mathbb{R}^q$, let A_i^j be the components of A so that the jth component of Ax is $A_i^j x^i$ summed over i. The chain rule takes the form:

$$
(f')_i^j = \frac{\partial y^j}{\partial x^i}, \quad (g')_j^k = \frac{\partial z^k}{\partial y^j}, \quad (h')_i^k = \frac{\partial z^k}{\partial x^i} = \sum_{j=1}^n (g')_j^k (f')_i^j = \sum_{j=1}^n \frac{\partial z^k}{\partial y^j}\frac{\partial y^j}{\partial x^i} \,.
$$

1.4   THE INVERSE AND IMPLICIT FUNCTION THEOREMS

This section deals with the Inverse Function Theorem and with the Implicit Function Theorem. A good auxiliary reference would be Spivak [38]. The Implicit Function Theorem is also known as Dini's Theorem, owing to the seminal contribution of the Italian mathematician Ulisse Dini.

Ulisse Dini (1845–1918)

Recall that if A is a linear transformation from $\mathbb{R}^m$ to $\mathbb{R}^n$, we define the *operator norm* $\|A\|$ by setting

$$\|A\| := \sup_{0 \neq x \in \mathbb{R}^m} \frac{\|Ax\|}{\|x\|} < \infty.$$

We then have $\|Ax\| \leq \|A\| \cdot \|x\|$ for all $x \in \mathbb{R}^m$. Also, recall that in $\mathbb{R}^m$ there is a canonical notation for the inner product; thus, we set $x \cdot y = x^1 y^1 + \cdots + x^m y^m$ to define the inner product of $(x^1, \ldots, x^m)$ and $(y^1, \ldots y^m)$. We begin our discussion with:

Lemma 1.7   Let $\mathcal{O}$ be an open subset of $\mathbb{R}^m$. Let $F : \mathcal{O} \to \mathbb{R}^n$ be continuously differentiable.

1. Let C be a convex compact subset of $\mathcal{O}$. Let $\kappa := \sup_{P \in C} \|F'(P)\| < \infty$. Then

$$\|F(P_1) - F(P)\| \leq \kappa \|P_1 - P\|.$$

2. If F is real-valued and if $F'(P) \neq 0$, then F does not have a local minimum at P.

3. Suppose $n = m$ and that F' is invertible on $\mathcal{O}$.

    (a) Let $\mathcal{U}$ be an open set with compact closure in $\mathcal{O}$. Let

    $$\kappa := \sup_{P \in \bar{\mathcal{U}}} \|(F'(P))^{-1}\| < \infty.$$

    If $P \in \mathcal{U}$, then there exists $\delta = \delta(P, F) > 0$ so $\bar{B}_\delta(P) \subset \mathcal{U}$ and:

    i. If $P_1 \in B_\delta(P)$ and $P_2 \in B_\delta(P)$, then $\|P_1 - P_2\| \leq 2\kappa \|F(P_1) - F(P_2)\|$.

    ii. F is an injective map from $B_\delta(P)$ to $\mathbb{R}^n$.

    iii. There exists $\sigma > 0$ so $F(B_\delta(P))$ contains $B_\sigma(F(P))$.

    (b) F is an open map.

Proof. If $F(P_1) = F(P)$, the inequality of Assertion 1 is obvious. Consequently we may assume that $F(P_1) \neq F(P)$ and set $v = (F(P_1) - F(P))/\|F(P_1) - F(P)\|$; $\|v\| = 1$. Since the region is convex, we can consider the straight line segment $tP_1 + (1 - t)P$ from P to P_1. We use the

Fundamental Theorem of Calculus and the Chain Rule to see

$$\|F(P_1) - F(P)\| = v \cdot \{F(P_1) - F(P)\} = \int_0^1 \partial_t \{v \cdot (F(tP_1 + (1-t)P))\}\, dt$$

$$= \int_0^1 v \cdot \{F'(tP_1 + (1-t)P)(P_1 - P)\}\, dt .$$

We apply the Cauchy–Schwarz–Bunyakovsky inequality to establish Assertion 1 by estimating:

$$\|F(P_1) - F(P)\| \leq \int_0^1 |v \cdot \{F'(tP_1 + (1-t)P)(P_1 - P)\}|\, dt$$

$$\leq \int_0^1 \|v\|\, \|F'(tP_1 + (1-t)P)(P_1 - P)\|\, dt$$

$$\leq \|P_1 - P\| \int_0^1 \|F'(tP_1 + (1-t)P)\|\, dt \leq \kappa \|P_1 - P\| .$$

If F is real-valued and if $F'(P) \neq 0$, choose u so $F'(P)u \neq 0$. Let $g(t) := F(P + tu)$. By the Chain Rule, $g'(0) = F'(P)u \neq 0$. Thus g does not have a local minimum at 0. This shows that F does not have a local minimum at P which proves Assertion 2.

Let $n = m$ and let F' be invertible on $\mathcal{O}$. Let $\mathcal{U}$ be an open set with compact closure in $\mathcal{O}$. Fix $P \in \mathcal{U}$ and let $A = F'(P)$. Let $\epsilon_1 := (2\kappa)^{-1}$ and let $G(x) := F(x) - Ax$. Since G is continuously differentiable, since $G'(x) = F'(x) - A$, and since $G'(P) = 0$, we can choose $\delta > 0$ so $B_\delta(P) \subset \mathcal{O}$ and so $\|G'(x)\| < \epsilon_1$ on $B_\delta(P)$. Thus by Assertion 1, if $P_i \in B_\delta(P)$

$$\|(F(P_1) - F(P_2)) - A(P_1 - P_2)\| = \|G(P_1) - G(P_2)\| \leq \epsilon_1 \|P_1 - P_2\| . \tag{1.4.a}$$

Since A is invertible,

$$\|P_1 - P_2\| = \|A^{-1}(A(P_1 - P_2))\| \leq \kappa \|A(P_1 - P_2)\| . \tag{1.4.b}$$

We use Equation (1.4.a), Equation (1.4.b), and the triangle inequality to estimate:

$$\|A(P_1 - P_2)\| \leq \|F(P_1) - F(P_2)\| + \epsilon_1 \|P_1 - P_2\|$$

$$\leq \|F(P_1) - F(P_2)\| + \epsilon_1 \|A^{-1}\| \cdot \|A(P_1 - P_2)\|$$

$$\leq \|F(P_1) - F(P_2)\| + \tfrac{1}{2} \|A(P_1 - P_2)\| .$$

This shows

$$\tfrac{1}{2} \|A(P_1 - P_2)\| \leq \|F(P_1) - F(P_2)\| . \tag{1.4.c}$$

Assertion 3-a-i now follows from Equation (1.4.b) and Equation (1.4.c). Assertion 3-a-ii follows from Assertion 3-a-i. Let $\eta := \tfrac{1}{2}\delta$ and let

$$S_\eta(P) := \{P_1 \in \mathbb{R}^m : \|P - P_1\| = \eta\}$$

be the sphere of radius η about P. Since F is injective on $\bar{B}_\eta(P) \subset B_\delta(P)$, $F(P)$ does not belong to $F(S_\eta(P))$. Since a continuous function attains its minimum on a compact subset and since, by Lemma 1.1, $S_\eta(P)$ is compact, we can choose $\sigma > 0$ so

$$\|F(P_1) - F(P)\| \geq 3\sigma \quad \text{for} \quad P_1 \in S_\eta(P).$$

Let $Q_1 \in B_\sigma(F(P))$. Let $f(x) := \|Q_1 - F(x)\|^2$. We use the triangle inequality to see

$$\|F(P_1) - Q_1\| \geq \|F(P_1) - F(P)\| - \|F(P) - Q_1\| \geq 3\sigma - \sigma = 2\sigma \quad \text{for} \quad P_1 \in S_\eta(P).$$

Since $\|F(P) - Q_1\| \leq \sigma$, f does not attain its minimum on $S_\eta(P)$. Because a continuous function attains its minimum on a compact set, we can choose $P_1 \in \bar{B}_\eta(P)$ so $f(P_1)$ is minimal. As noted above, $P_1 \notin S_\eta(P)$ so P_1 is in the open ball $B_\eta(P)$. By Assertion 2, $f'(P_1) = 0$. Let $(x^1, \ldots, x^m)$ be the usual coordinates on $\mathbb{R}^m$. We express

$$f(x) = \sum_{i=1}^m (F^i(x) - Q_1^i)^2 \quad \text{so} \quad \partial_{x^j} f(x) = 2 \sum_{i=1}^m (\partial_{x^j} F^i(x))(F^i(x) - Q_1^i).$$

Setting $\partial_{x^j} f(P_1) = 0$ for $1 \leq j \leq m$ gives rise to the equations:

$$0 = \sum_{i=1}^m (\partial_{x^j} F^i(P_1))(F^i(P_1) - Q_1^i) \quad \text{for} \quad 1 \leq j \leq m$$

which can be written in matrix form as: $0 = F'(P_1)(F(P_1) - Q_1)$. Since, by assumption, $F'(P_1)$ is invertible we have $F(P_1) = Q_1$. This proves Assertion 3-a-iii. Thus $F(\mathcal{O})$ contains an open neighborhood of $F(P)$. Since P was arbitrary, $F(\mathcal{O})$ is open. Applying the same argument to an arbitrary open subset of $\mathcal{O}$ then establishes Assertion 3-b. $\square$

Theorem 1.8 (Inverse Function Theorem). *Let $\mathcal{O}$ be an open subset of $\mathbb{R}^m$ and let F be a continuously differentiable function mapping $\mathcal{O}$ to $\mathbb{R}^m$ with F' invertible on $\mathcal{O}$. Fix $P \in \mathcal{O}$. There exists an open neighborhood $\mathcal{U}$ of P with $\mathcal{U} \subset \mathcal{O}$ and there exists an open neighborhood $\mathcal{V} \subset F(\mathcal{O})$ of $Q = F(P)$ so $F : \mathcal{U} \to \mathcal{V}$ is bijective. Let G be the inverse map. Then G is continuously differentiable and $G'(Q) = \{F'(G(Q))\}^{-1}$.*

Proof. The Inverse Function Theorem is a local result. Let $\mathcal{O}_1$ be an open neighborhood of P which has compact closure in $\mathcal{O}$. Because $(F')^{-1}$ is continuous on $\bar{\mathcal{O}}_1$,

$$\kappa := \max_{P_1 \in \bar{\mathcal{O}}_1} \|(F'(P_1))^{-1}\|$$

is well defined. Thus by replacing $\mathcal{O}$ by $\mathcal{O}_1$ if necessary, we may assume without loss of generality that $\|(F')^{-1}\| \leq \kappa$ on all of $\mathcal{O}$. Let $\mathcal{U} = B_\delta(P) \subset \mathcal{O}$ be given by Lemma 1.7. By Lemma 1.7, $\mathcal{V} := F(\mathcal{U})$ is open and F is a bijective map from $\mathcal{U}$ to $\mathcal{V}$. We restate the estimate

$$\|P_1 - P_2\| \leq 2\kappa \|F(P_1) - F(P_2)\|$$

in the form:

$$\|G(Q_1) - G(Q_2)\| \leq 2\kappa \|Q_1 - Q_2\| \quad \text{for} \quad Q_i \in \mathcal{V}. \tag{1.4.d}$$

This proves G is continuous. We use the following equations to define $\mathcal{E}_F$ and $\mathcal{E}_G$:

$$F(P_1) = F(P_2) + F'(P_2)(P_1 - P_2) + \mathcal{E}_F(P_1, P_2) \quad \text{for} \quad P_i \in \mathcal{U}, \tag{1.4.e}$$
$$G(Q_1) = G(Q_2) + \{F'(G(Q_2))\}^{-1}(Q_1 - Q_2) + \mathcal{E}_G(Q_1, Q_2) \quad \text{for} \quad Q_i \in \mathcal{V}. \tag{1.4.f}$$

We set $P_1 = G(Q_1)$ and $P_2 = G(Q_2)$ and use Equation (1.4.e) to see:

$$Q_1 = Q_2 + F'(G(Q_2))(G(Q_1) - G(Q_2)) + \mathcal{E}_F(G(Q_1), G(Q_2)) \quad \text{for} \quad Q_i \in \mathcal{V}. \tag{1.4.g}$$

We multiply Equation (1.4.g) by $\{F'(G(Q_2))\}^{-1}$ and perform some algebraic rearrangements to see

$$G(Q_1) - G(Q_2) = \{F'(G(Q_2))\}^{-1}\{(Q_1 - Q_2) - \mathcal{E}_F(G(Q_1), G(Q_2))\}. \tag{1.4.h}$$

We compare Equation (1.4.f) with Equation (1.4.h) to see:

$$\mathcal{E}_G(Q_1, Q_2) = -\{F'(G(Q_2))\}^{-1}\mathcal{E}_F(G(Q_1), G(Q_2)). \tag{1.4.i}$$

Let $\epsilon > 0$ be given. Fix $Q_1 \in \mathcal{V}$ and let $P_1 = G(Q_1) \in \mathcal{U}$. Since F is differentiable at P_1, we may choose $\delta_1 = \delta_1(Q_1) > 0$ so $\|P_1 - P_2\| < \delta_1$ implies

$$\|\mathcal{E}_F(P_1, P_2)\| \leq \frac{\epsilon}{2\kappa^2}\|P_1 - P_2\|.$$

Since G is continuous, we may find $\delta > 0$ so $\|Q_1 - Q_2\| < \delta$ implies $\|G(Q_1) - G(Q_2)\| < \delta_1$. Since $\|(F')^{-1}\| \leq \kappa$ on $\mathcal{O}$, Equation (1.4.d), and Equation (1.4.i) imply:

$$
\begin{aligned}
\|\mathcal{E}_G(Q_1, Q_2)\| &\leq \|\{F'(G(Q_2))\}^{-1}\| \cdot \|\mathcal{E}_F(G(Q_1), G(Q_2))\| \\
&\leq \kappa \frac{\epsilon}{2\kappa^2}\|G(Q_1) - G(Q_2)\| \leq 2\kappa^2 \frac{\epsilon}{2\kappa^2}\|Q_1 - Q_2\| \\
&\leq \epsilon\|Q_1 - Q_2\|.
\end{aligned}
$$

It now follows $G'(Q_1) = \{F'(G(Q_1))\}^{-1}$ for $Q_1 \in \mathcal{V}$. Since F is continuously differentiable, we can use Cramer's rule to see $(F')^{-1}$ is continuous. Since G is continuous, it follows that G is continuously differentiable. $\qquad \square$

1.4.1 REMARK. If F is *smooth*, i.e., has continuous partial derivatives of all orders, then we can use the equation $G' = (F')^{-1} \circ G$ to conclude that the local inverse G is smooth as well; one simply differentiates this equation recursively and uses the Chain Rule.

If $\mathcal{O}$ and $\mathcal{U}$ are open subsets of $\mathbb{R}^m$, if F is a continuous map from $\mathcal{O}$ to $\mathcal{U}$, and if F^{-1} is continuous, then F is said to be a *homeomorphism*. If F is smooth and if F^{-1} is smooth, then F is said to be a *diffeomorphism*. It follows from the Chain Rule that the composition of diffeomorphisms is a diffeomorphism. It follows from the Inverse Function Theorem that the inverse of a diffeomorphism is a diffeomorphism. We now present several examples of canonical coordinate systems.

1.4.2 REMARK. We consider the function

$$f(x) = \left\{ \begin{array}{ll} 0 & \text{if } x = 0 \\ \frac{x}{2} + x^2 \sin(x^{-1}) & \text{if } x \neq 0 \end{array} \right\} .$$

This function is differentiable everywhere and $f'(0) = \frac{1}{2}$. The derivative is not continuous at $x = 0$. This function is not injective on any neighborhood of 0; thus the assumption that f has continuous partial derivatives is crucial in the Inverse Function Theorem.

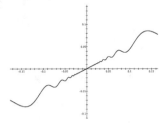

1.4.3 POLAR COORDINATES. Let $F(r, \theta) := (r \cos(\theta), r \sin(\theta))$, i.e.,

$$x = r \cos(\theta), \quad y = r \sin(\theta) \quad \text{for} \quad r > 0, \ \theta \in \mathbb{R} .$$

Then $|\det(F')| = r$ and a local inverse near $x = 1$ and $y = 0$ is given by

$$r = (x^2 + y^2)^{\frac{1}{2}} \quad \text{and} \quad \theta = \arctan\left(\frac{y}{x}\right) .$$

The natural domain of the local inverse is the right half plane $y > 0$ where we choose the branch of the arctan function to have range $(-\frac{\pi}{2}, \frac{\pi}{2})$. This is not globally invertible as we have the relation $F(r, \theta) = F(r, \theta + 2\pi)$. We say that this is an *admissible change of coordinates* since we can express (r, θ) in terms of (x, y) and (x, y) in terms of (r, θ) at least locally. This permits us to express certain curves very simply. For example, one has:

Archimedes spiral
$r = \theta$

Region bounded by
Cardioid
$r = \sin(\theta) - 1$

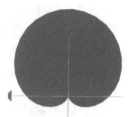

1.4.4 CYLINDRICAL COORDINATES. Let $F(r, \theta, z) := (r \cos(\theta), r \sin(\theta), z)$, i.e.,

$$x = r \cos(\theta), \quad y = r \sin(\theta), \quad z = z \quad \text{for} \quad r > 0, \ (\theta, z) \in \mathbb{R}^2 .$$

Then $|\det(F')| = r$ so again this is an admissible change of coordinates, i.e., we can write (r, θ, z) in terms of (x, y, z), locally. Again, the inverse is not globally injective but can be used to provide a local parametrization of $\mathbb{R}^3$ minus the z-axis. Thus for example,

$z = r$ yields the cone

1.4.5 SPHERICAL COORDINATES. We let

$$x = r\cos(\theta)\sin(\phi), \quad y = r\sin(\theta)\sin(\phi), \quad z = r\cos(\phi) \quad \text{for} \quad r > 0, \; \theta \in \mathbb{R}, \; 0 < \phi < \pi \, .$$

Since $|\det(F')| = r^2\sin(\phi)$, F is locally injective; as with polar and cylindrical coordinates it is not globally injective but F^{-1} is well-defined and smooth locally. The parameter r measures the distance to the origin, the parameter θ is the angle in the xy plane, and the parameter ϕ is the angle down from the z-axis. Again, we get wonderful surfaces. For example,

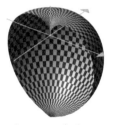

$r = \phi$ and $0 \le \theta \le \pi$ yields:

The Inverse Function Theorem can be regarded as a change of coordinates. We use this point of view to establish the following result; note the assumption of differentiability in the following result is in fact not needed and methods of algebraic topology (Invariance of Domain) can be used to prove it in the C^0 category.

Theorem 1.9   *Let $\mathcal{O}$ be an open non-empty subset of $\mathbb{R}^m$ and let $H : \mathcal{O} \to \mathbb{R}^n$ be continuously differentiable. If H is injective, then $n \ge m$.*

Proof. We suppose the theorem fails. By padding out the coordinates if necessary, we may assume $n = m - 1$. Choose a counter example where $m \ge 2$ is minimal. If $\partial_{x^1} H \equiv 0$, then

$$H(x + \delta e_1) - H(x) = \int_0^\delta \partial_t H(x + te_1)dt = \int_0^\delta \partial_{x^1} H(x + te_1)dt = 0 \, .$$

This is impossible as H is injective. Thus there exists a point $P \in \mathcal{O}$ so $\partial_{x^1} H(P) \ne 0$. By reordering the components if necessary, we may assume $\partial_{x^1} H_1(P) \ne 0$. Let

$$F(x) := (H_1(x), x^2, \ldots, x^m) \, .$$

Then $\det(F') = \partial_{x^1} H_1$. Thus by the Inverse Function Theorem, F is invertible near P. We set $\tilde{H}(y^1,\ldots,y^m) := H(F^{-1}(y^1,\ldots,y^m))$; this is defined near $F(P)$ and is continuously differentiable. Since the composition of injective functions is injective, $\tilde{H}$ is injective. Tracing through the definitions yields

$$\tilde{H}_1(y^1,\ldots,y^m) = H_1(F^{-1}(y^1,\ldots,y^m)) = F_1(F^{-1}(y^1,\ldots,y^m)) = y^1 \quad \text{and}$$
$$\tilde{H}(y^1,\ldots,y^m) = (y^1, \tilde{H}_2(y^2,\ldots,y^m),\ldots,\tilde{H}_{m-1}(y^1,\ldots,y^m)).$$

If $m = 2$, then $\tilde{H}(y^1, y^2) = y^1$ and this is not an injective map. Thus $m \geq 3$ and fixing y^1 gives an injective map from $\mathbb{R}^{m-1}$ to $\mathbb{R}^{m-2}$

$$\tilde{F}(y^2,\ldots,y^m) := (\tilde{H}_2(y^1, y^2,\ldots,y^m),\ldots,\tilde{H}_{m-1}(y^1, y^2,\ldots,y^m)).$$

This gives a counter example in dimension $m - 1$ contradicting the minimality of m.   □

What is going on, of course, is that one is gradually changing the coordinate system to put H into a particularly simple form. This is a typical application of the Inverse Function Theorem. We will also use the Inverse Function Theorem to prove the Implicit Function Theorem; this permits one to solve equations to determine certain variables (called the dependent variables) in terms of other variables (the independent variables). This is best motivated by examples so we shall first present several examples and then summarize our algorithm with the Implicit Function Theorem.

1.4.6 EXAMPLE. Consider the equation

$$x^5 + x + y^5 + y = 4. \tag{1.4.j}$$

Suppose $P = (x_0, y_0)$ solves Equation (1.4.j); for example we could take $P = (1, 1)$. We let x be the *independent variable* and y the *dependent variable*. We try to solve this equation for $y = y(x)$ near the point P. Let $F(x, y) := (x, x^5 + x + y^5 + y)$. Here and subsequently let "$\star$" be a term which is not of interest. We have:

$$\det(F'(P))(x_0, y_0) = \det \begin{pmatrix} 1 & 0 \\ \star & 5y_0^4 + 1 \end{pmatrix} = 5y_0^4 + 1 \neq 0.$$

This is invertible. If $F(x, y) = (u, v)$, then the inverse function $G(u, v) = (x, y)$ has the property that

$$G_1(x, v) = x \quad \text{and} \quad y(x) := G_2(x, 4)$$

satisfies $y(x_0) = y_0$ and $x^5 + x + y^5(x) + y(x) = 4$. In other words, we have found the unique solution to Equation (1.4.j) where $y(x_0) = y_0$ and where $y(x)$ is close to y_0. The map $x \to y(x)$ is continuously differentiable. We *implicitly differentiate* Equation (1.4.j) to see:

$$y'(x) = -\frac{5x^4 + 1}{5y^4(x) + 1}. \tag{1.4.k}$$

The Inverse Function Theorem implies $y \in C^1$. We use Equation (1.4.k) to see $y' \in C^1$ so $y \in C^2$ has two continuous derivatives. We continue in this fashion to see that in fact $y \in C^\infty$. (In fact y is real analytic).

The curve $x^5 + x + \{y(x)\}^5 + y(x) = 4$.

1.4.7 EXAMPLE. Consider the equations:

$$x^2 + y^4 + z^8 = 3 \quad \text{and} \quad x^4 + y^6 + z^4 = 3.$$

Let x be the independent variable and (y, z) the dependent variables. We wish to solve $y = y(x)$ and $z = z(x)$ near the point $(1, 1, 1)$. We set

$$F(x, y, z) = (x, x^2 + y^4 + z^8, x^4 + y^6 + z^4).$$

The Jacobian is

$$F'(1, 1, 1) = \begin{pmatrix} 1 & 0 & 0 \\ \star & 4 & 8 \\ \star & 6 & 4 \end{pmatrix}.$$

Since $\det(F'(1, 1, 1)) \neq 0$, we can apply the Inverse Function Theorem. Let G be the inverse. Then $y(x) = G_2(x, 3, 3)$ and $z(x) = G_3(x, 3, 3)$ satisfies the equations $x^2 + y^4 + z^8 = 3$ and $x^4 + y^6 + z^4 = 3$. We differentiate to see

$$2x + 4y^3 \partial_x y + 8z^7 \partial_x z = 0 \quad \text{and} \quad 4x^3 + 6y^5 \partial_x y + 4z^3 \partial_x z = 0.$$

Expressing these equations in matrix notation and then inverting yields:

$$\begin{pmatrix} 4y^3 & 8z^7 \\ 6y^5 & 4z^3 \end{pmatrix} \begin{pmatrix} \frac{\partial y}{\partial x} \\ \frac{\partial z}{\partial x} \end{pmatrix} = \begin{pmatrix} -2x \\ -4x^3 \end{pmatrix}$$

$$\begin{pmatrix} \frac{\partial y}{\partial x} \\ \frac{\partial z}{\partial x} \end{pmatrix} = \begin{pmatrix} 4y^3 & 8z^7 \\ 6y^5 & 4z^3 \end{pmatrix}^{-1} \begin{pmatrix} -2x \\ -4x^3 \end{pmatrix}.$$

The right hand side is a C^1 function of x. Thus the left hand side is C^1. This implies $y(x)$ and $z(x)$ are C^2. We proceed inductively to see these functions are in fact C^∞.

$x^2 + y^4 + z^8 = 3$ and $x^4 + y^6 + z^4 = 3$ yields:

1.4.8 EXAMPLE. Consider the equations:

$$x^5 + 5x + y^7 + 7y + u^6 + v^8 = 16,$$
$$x^3 + 2x + y^5 + 2y + u^4 v^6 = 7.$$

(1.4.1)

Let $P = (1, 1, 1, 1)$; this solves these equations. We first let $\{x, y\}$ be the independent variables and $\{u, v\}$ be the dependent variables and try to solve for (u, v) in terms of (x, y). Let

$$F(x, y, u, v) = (x, y, x^5 + 5x + y^7 + 7y + u^6 + v^8, x^3 + 2x + y^5 + 2y + u^4 v^6).$$

We have:

$$F'(1, 1, 1, 1) = \begin{pmatrix} 1 & 0 & 0 & 0 \\ 0 & 1 & 0 & 0 \\ * & * & 6u^5 & 8v^7 \\ * & * & 4u^3 v^6 & 6u^4 v^5 \end{pmatrix}\Bigg|_{(1,1,1,1)} = \begin{pmatrix} 1 & 0 & 0 & 0 \\ 0 & 1 & 0 & 0 \\ * & * & 6 & 8 \\ * & * & 4 & 6 \end{pmatrix}.$$

Thus $\det(F'(1, 1, 1, 1)) = 36 - 32 \neq 0$. The Inverse Function Theorem shows $G := F^{-1}$ is well defined and C^1. Then $x = G_1(x, \star, \star, \star)$ and $y = G_2(\star, y, \star, \star)$. If we set

$$u(x, y) := G_3(x, y, 16, 7) \quad \text{and} \quad v(x, y) := G_4(x, y, 16, 7),$$

then $(x, y, u(x, y), v(x, y))$ is the unique solution to Equation (1.4.1) that is near $(1, 1, 1, 1)$. The pair (u, v) are C^1 with $u(1, 1) = 1$ and $v(1, 1) = 1$. If we differentiate implicitly, we obtain the equations:

$$\begin{pmatrix} 6u^5 & 8v^7 \\ 4u^3 v^6 & 6u^4 v^5 \end{pmatrix} \begin{pmatrix} \frac{\partial u}{\partial x} & \frac{\partial u}{\partial y} \\ \frac{\partial v}{\partial x} & \frac{\partial v}{\partial y} \end{pmatrix} = - \begin{pmatrix} 5x^4 + 5 & 7y^6 + 7 \\ 3x^2 + 2 & 5y^4 + 2 \end{pmatrix}.$$

This yields

$$\begin{pmatrix} \frac{\partial u}{\partial x} & \frac{\partial u}{\partial y} \\ \frac{\partial v}{\partial x} & \frac{\partial v}{\partial y} \end{pmatrix} = - \begin{pmatrix} 6u^5 & 8v^7 \\ 4u^3 v^6 & 6u^4 v^5 \end{pmatrix}^{-1} \begin{pmatrix} 5x^4 + 5 & 7y^6 + 7 \\ 3x^2 + 2 & 5y^4 + 2 \end{pmatrix}.$$

If the right hand side is C^k for $k \geq 1$, then the left hand side is C^k. This implies the right hand side is C^{k+1} and hence the functions $u(\cdot, \cdot)$ and $v(\cdot, \cdot)$ are C^∞.

1.4.9 EXAMPLE. We consider the same relations as those given in Equation (1.4.1) but instead let (u, v) be the independent variables, we let (x, y) be the dependent variables, and we try to solve for $x = x(u, v)$ and $y = y(u, v)$. We now set

$$F(x, y, u, v) = (u, v, x^5 + 5x + y^7 + 7y + u^6 + v^8, x^3 + 2x + y^5 + 2y + u^4 v^6).$$

The Jacobian is given by

$$F'(1, 1, 1, 1) = \begin{pmatrix} 1 & 0 & 0 & 0 \\ 0 & 1 & 0 & 0 \\ \star & \star & 5x^4 + 5 & 7y^6 + 7 \\ \star & \star & 3x^2 + 2 & 5y^4 + 2 \end{pmatrix}\Bigg|_{(1,1,1,1)} = \begin{pmatrix} 1 & 0 & 0 & 0 \\ 0 & 1 & 0 & 0 \\ \star & \star & 10 & 14 \\ \star & \star & 5 & 7 \end{pmatrix}.$$

Since $\det(F'(1, 1, 1, 1)) = 0$, we cannot employ the Inverse Function Theorem. Nevertheless, we still assume we could somehow find C^1 solutions so $x = x(u, v)$ and $y = y(u, v)$. We differentiate implicitly to get:

$$\begin{pmatrix} 5x^4 + 5 & 7y^6 + 7 \\ 3x^2 + 2 & 5y^4 + 2 \end{pmatrix} \begin{pmatrix} \frac{\partial x}{\partial u} & \frac{\partial x}{\partial v} \\ \frac{\partial y}{\partial u} & \frac{\partial y}{\partial v} \end{pmatrix} = -\begin{pmatrix} 6u^5 & 8v^7 \\ 4u^3 v^6 & 6u^4 v^5 \end{pmatrix}.$$

At the point $(1, 1, 1, 1)$ this would yield:

$$\begin{pmatrix} 10 & 14 \\ 5 & 7 \end{pmatrix} \begin{pmatrix} \frac{\partial x}{\partial u} & \frac{\partial x}{\partial v} \\ \frac{\partial y}{\partial u} & \frac{\partial y}{\partial v} \end{pmatrix} = -\begin{pmatrix} 6 & 8 \\ 4 & 6 \end{pmatrix}.$$

The determinant of the left hand side is zero; the determinant of the right hand side is -4. This is not possible and thus the equations cannot be solved in a C^1 fashion for (x, y) as a function of (u, v) near $(1, 1, 1, 1)$.

1.4.10 IMPLICIT FUNCTION THEOREM. With these examples in mind, we can now establish the following result:

Theorem 1.10   (Implicit Function Theorem).   *Let $\mathcal{U}$ be an open subset of $\mathbb{R}^a$ and let $\mathcal{V}$ be an open subset of $\mathbb{R}^b$. Let $x = (x^1, \dots, x^a)$ belong to $\mathcal{U}$ and let $y = (y^1, \dots, y^b)$ belong to $\mathcal{V}$; x are the independent variables and y are the dependent variables. Let $H = H(x, y)$ be a continuously differentiable map from $\mathcal{U} \times \mathcal{V}$ to $\mathbb{R}^b$. Let $(P, Q) \in \mathcal{U} \times \mathcal{V}$. Let A be the $b \times b$ matrix $(\partial_{y^i} H^j)(P, Q)$. Assume $\det(A) \neq 0$. There exist open neighborhoods $\mathcal{U}_1$ of P and $\mathcal{V}_1$ of Q so that:*

1. *If $x \in \mathcal{U}_1$, there exists a unique $y \in \mathcal{V}_1$ so $H(x, y) = H(P, Q)$.*

2. *The map $f : x \to y(x)$ is a C^1 map from $\mathcal{U}_1$ to $\mathcal{V}_1$ and*

$$\left\{ \sum_i \frac{\partial H^j}{\partial y^i} \frac{\partial f^i}{\partial x^k} + \frac{\partial H^j}{\partial x^k} \right\} (x, f(x)) = 0 \quad for \quad 1 \le k \le a, 1 \le j \le b.$$

3. Let $k \geq 2$. If H is C^k, then f is C^k.

Proof. As before, we define an auxiliary function $F(x, y) := (x, H(x, y))$ from $\mathcal{U} \times \mathcal{V}$ to $\mathbb{R}^{a+b}$ and compute:

$$F'(P, Q) = \begin{pmatrix} \mathrm{Id}_a & 0 \\ \star & A \end{pmatrix}.$$

Since $\det(F'(P, Q)) = \det(A) \neq 0$, we have a local inverse $G = (G_1, G_2)$ where G_1 takes values in $\mathbb{R}^a$ and G_2 takes values in $\mathbb{R}^b$. We have $G_1(x, \star) = x$. Let $f(x) = G_2(x, H(P, Q))$. Then $H(x, f(x)) = H(P, Q)$. The Assertion 1 of the Lemma follows; we use the Chain Rule to establish Assertion 2. We invert the equation of Assertion 2 to express $\frac{\partial y}{\partial x}$ in terms of $(\frac{\partial H}{\partial y})^{-1}$ and $\frac{\partial f}{\partial x}$. We differentiate recursively and apply the Chain Rule once again to obtain Assertion 3. $\square$

We note that if H is C^∞, then f is C^∞. With a bit more work, one can show that if H is real analytic (i.e., given by a convergent Taylor series), then f is real analytic. In Example 1.4.9, we gave an example where the hypotheses of the Implicit Function Theorem failed and where one cannot in fact solve the equations. On the other hand, if one considers the quadratic equation $x^2 - 2xy + y^2 = 0$, then again the hypotheses of the Implicit Function Theorem fail since all derivatives vanish at $x = y = 0$. This equation is equivalent to the equation $(x - y)^2 = 0$, i.e., $y = x$. Thus when the hypotheses of the Implicit Function Theorem fail, little can be said in general although there are certain methods which pertain and which these two examples illustrate.

1.5 THE RIEMANN INTEGRAL

In this section, we shall establish the results concerning the Riemann integral needed subsequently; this integral was created by the German mathematician Georg Friedrich Bernhard Riemann.

G. Riemann (1826–1866)

The theory closely parallels that of the 1-dimensional case and a good auxiliary reference would be Rudin [36]. A *rectangle* $R \subset \mathbb{R}^m$ is the Cartesian product

$$R = [a_1, b_1] \times \cdots \times [a_m, b_m] \quad \text{for} \quad a_1 < b_1, \ldots, a_m < b_m.$$

We do not consider degenerate rectangles where $a_i = b_i$ for some i. The *volume* of R is then defined to be:

$$\text{vol}(R) := (b_1 - a_1) \cdot \dots \cdot (b_m - a_m) > 0. \tag{1.5.a}$$

A *partition* of an interval $[a, b] \subset \mathbb{R}$ is simply a collection of intermediate points

$$\mathcal{P} := \{a = c_0 < \dots < c_v = b\}.$$

By an abuse of notation we may identify $\mathcal{P}$ with a dissection of $[a, b]$ of the form:

$$\mathcal{P} := \{[a, b] = [c_0, c_1] \cup [c_1, c_2] \cup \dots \cup [c_{v-1}, c_v]\}.$$

More generally, a partition $\mathcal{P}$ of a rectangle $R \subset \mathbb{R}^m$ is a collection

$$\mathcal{P} = \{\{a_1 = c_0^1 < \dots < c_{v_1}^1 = b_1\}, \dots, \{a_m = c_0^m < \dots < c_{v_m}^m = b_m\}\}$$

which we identify with a dissection of R of the form:

$$\mathcal{P} = \{R = \cup_{1 \leq i_1 \leq v_1, \dots, 1 \leq i_m \leq v_m} \{[c_{i_1-1}^1, c_{i_1}^1] \times \dots \times [c_{i_m-1}^m, c_{i_m}^m]\}\}.$$

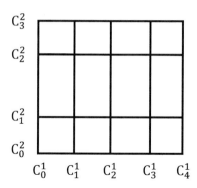

Lemma 1.11   If $\mathcal{P}$ is a partition of a rectangle $R \subset \mathbb{R}^m$, then $\sum_{P \in \mathcal{P}} \text{vol}(P) = \text{vol}(R)$.

Proof. If $m = 1$, this is nothing but the fact that we have a telescoping series

$$\sum_{P \in \mathcal{P}} \text{vol}(P) = (c_1 - c_0) + \dots + (c_v - c_{v-1}) = c_v - c_0 = b - a = \text{vol}(R).$$

In higher dimensions, we use this argument together with the distributive law:

$$\sum_{P \in \mathcal{P}} \text{vol}(P) = \sum_{i_1=1}^{v_1} \dots \sum_{i_m=1}^{v_m} (c_{i_1}^1 - c_{i_1-1}^1) \cdot \dots \cdot (c_{i_m}^m - c_{i_m-1}^m)$$

$$= \left\{ (c_1^1 - c_0^1) + \dots + (c_{v_1}^1 - c_{v_1-1}^1) \right\} \cdot \dots \cdot \left\{ (c_1^m - c_0^m) + \dots + (c_{v_m}^m - c_{v_m-1}^m) \right\}$$

$$= (b_1 - a_1) \cdot \dots \cdot (b_m - a_m) = \text{vol}(R). \qquad \square$$

Let f be a bounded real-valued function on a rectangle $R \subset \mathbb{R}$. If $P \subset R$, set:

$$M(f, P) := \sup_{x \in P} f(x), \qquad m(f, P) := \inf_{x \in P} f(x).$$

Clearly $m(f, P) \le M(f, P)$. If $\mathcal{P}$ is a partition of R, we define the *lower sum* $L(f, \mathcal{P})$ and the *upper sum* $U(f, \mathcal{P})$ to be:

$$L(f, \mathcal{P}) := \sum_{P \in \mathcal{P}} m(f, P) \operatorname{vol}(P) \quad \text{and} \quad U(f, \mathcal{P}) := \sum_{P \in \mathcal{P}} M(f, P) \operatorname{vol}(P).$$

We may estimate:

$$
\begin{aligned}
\operatorname{vol}(R)m(f, R) = \sum_{P \in \mathcal{P}} \operatorname{vol}(P) \inf_{x \in R} f(x) &\le \sum_{P \in \mathcal{P}} \operatorname{vol}(P) \inf_{x \in P} f(x) = L(f, P), \\
\operatorname{vol}(R)M(f, R) = \sum_{P \in \mathcal{P}} \operatorname{vol}(P) \sup_{x \in R} f(x) &\ge \sum_{P \in \mathcal{P}} \operatorname{vol}(P) \sup_{x \in P} f(x) = U(f, P).
\end{aligned}
\tag{1.5.b}
$$

A partition $\mathcal{Q}$ of R is said to be a *refinement* of a partition $\mathcal{P}$ of R if all the rectangles of $\mathcal{Q}$ are contained in some rectangle of $\mathcal{P}$. Since $\mathcal{Q}$ induces a partition of every rectangle of $\mathcal{P}$, the inequalities of Equation (1.5.b) can be summed over the rectangles $P \in \mathcal{P}$ to see:

$$L(f, \mathcal{P}) \le L(f, \mathcal{Q}) \le U(f, \mathcal{Q}) \le U(f, \mathcal{P}).$$

Given any two partitions $\mathcal{P}_1$ and $\mathcal{P}_2$, we can form a common refinement $\mathcal{Q}$ and conclude:

$$L(f, \mathcal{P}_1) \le L(f, \mathcal{Q}) \le U(f, \mathcal{Q}) \le U(f, \mathcal{P}_2).$$

This inequality shows that $L(f, \cdot)$ is uniformly bounded from above and $U(f, \cdot)$ is uniformly bounded from below. We therefore define the *lower integral* and the *upper integral* by setting:

$$\underline{\int_R} f := \sup_{\mathcal{P} \text{ is a partition of } R} L(f, \mathcal{P}) \quad \text{and} \quad \overline{\int_R} f := \inf_{\mathcal{P} \text{ is a partition of } R} U(f, \mathcal{P}).$$

We then have $\underline{\int_R} f \le \overline{\int_R} f$. The following observation is immediate:

Lemma 1.12   Let R be a rectangle in $\mathbb{R}^m$.

1. The following assertions are equivalent and if any is satisfied, then f is said to be *integrable* on R:

   (a) $\underline{\int_R} f = \overline{\int_R} f$.

   (b) Given $\epsilon > 0$, there exists a partition $\mathcal{P}$ of R so $U(f, \mathcal{P}) \le L(f, \mathcal{P}) + \epsilon$.

2. Let f be integrable on R. Set $\int_R f := \underline{\int}_R f = \overline{\int}_R f$.

   (a) If $\mathcal{P}$ is a partition of R satisfying Assertion 1-b, then $\left|\int_R f - L(f,\mathcal{P})\right| \le \epsilon$ and $\left|\int_R f - U(f,\mathcal{P})\right| \le \epsilon$.

   (b) Let P be a rectangle which is contained in R. Then $f|_P$ is integrable on P.

   (c) Let $\mathcal{P}$ be a partition of R. Then $\int_R f = \sum_{P\in\mathcal{P}} \int_P f|_P$.

   (d) If $c \in \mathbb{R}$, then cf is integrable on R and $\int_R cf = c\int_R f$.

3. Let f and g be integrable on R.

   (a) $f + g$ is integrable on R and $\int_R(f + g) = \int_R f + \int_R g$.

   (b) If $f(x) \le g(x)$ for all $x \in R$, then $\int_R f \le \int_R g$.

We will sometimes simply write $\int f$ when region of integration is clear and introducing additional notation would only complicate matters unnecessarily. If $R = [a, b] \subset \mathbb{R}$, then we will also use the standard notation $\int_a^b f(x)dx$.

The following example is instructive.

1.5.1 EXAMPLE. Let $\{r_n\}$ be an enumeration of the rationals in $[0, 1]$. Let

$$f(x) := \begin{cases} \frac{1}{n} & \text{if } x = r_n \text{ for some n} \\ 0 & \text{if } x \ne r_n \text{ for all } n \end{cases}. \quad \text{Graph:}$$

Fix n and let $\mathcal{P}_n$ be the partition of $[0, 1]$ into n^2 intervals of length $\frac{1}{n^2}$;

$$\mathcal{P}_n = \left\{0, \frac{1}{n^2}, \frac{2}{n^2}, \ldots, \frac{n^2-1}{n^2}, 1\right\}.$$

Let $\mathcal{P}_n^1$ be the collection of rectangles which contain r_k for some $k \le n$ and let $\mathcal{P}_n^2$ be the remaining rectangles. Since any rational number is in at most two of the intervals, $|\mathcal{P}_n^1| \le 2n$. Furthermore, $|f(x)| \le \frac{1}{n}$ for any $x \in P \in \mathcal{P}_n^2$. Since $0 \le f \le 1$, we may estimate:

$$0 \le L(f,\mathcal{P}_n) \le U(f,\mathcal{P}_n) \le \sum_{P\in\mathcal{P}_n^1} 1 \cdot \text{vol}(P) + \sum_{P\in\mathcal{P}_n^2} \frac{1}{n^2}\text{vol}(P) \le \frac{2n}{n^2} + \frac{1}{n^2}.$$

Since this tends to 0 as $n \to \infty$, f is integrable and $\int_{[0,1]} f = 0$. Note that f is discontinuous at every rational point of $[0, 1]$ and continuous at every irrational point of $[0, 1]$.

Let f be a bounded real-valued function on a rectangle $R \subset \mathbb{R}^m$. We have used closed rectangles to define $M(f, P)$ and $m(f, P)$. Occasionally, it is convenient to use open rectangles. Let $\mathcal{P}$ be a partition of R. Set:

$$\check{m}(f, P) := \inf_{x \in \text{int}(P)} f(x), \quad \check{L}(f, \mathcal{P}) := \sum_{P \in \mathcal{P}} \text{vol}(P) \check{m}(f, P),$$
$$\check{M}(f, P) := \sup_{x \in \text{int}(P)} f(x), \quad \check{U}(f, \mathcal{P}) := \sum_{P \in \mathcal{P}} \text{vol}(P) \check{M}(f, P).$$

Lemma 1.13 Let f be a bounded real-valued function on a rectangle $R \subset \mathbb{R}^m$. Let $\epsilon > 0$ be given and let $\mathcal{P}$ be a partition of R.

1. There exists a rectangle Q contained in the interior of R so $\text{vol}(R) - \text{vol}(Q) < \epsilon$.

2. $L(f, \mathcal{P}) \leq \check{L}(f, \mathcal{P}) \leq \check{U}(f, \mathcal{P}) \leq U(f, \mathcal{P})$.

3. There is a refinement $\mathcal{Q}$ of $\mathcal{P}$ so $\check{L}(f, \mathcal{P}) \leq L(f, \mathcal{Q}) + \epsilon$ and $U(f, \mathcal{Q}) - \epsilon \leq \check{U}(f, \mathcal{P})$.

Proof. Let $\delta < \frac{1}{2} \min_{1 \leq i \leq m} (b_i - a_i)$. Let $Q_\delta := [a_1 + \delta, b_1 - \delta] \times \cdots \times [a_m + \delta, b_m - \delta]$ be contained in the interior of R. Assertion 1 now follows since

$$\lim_{\delta \to 0} \text{vol}(Q_\delta) = \lim_{\delta \to 0} (b_1 - a_1 - 2\delta) \cdot \cdots \cdot (b_m - a_m - 2\delta) = \text{vol}(R).$$

Assertion 2 is immediate. Suppose first that the partition $\mathcal{P}$ consists only of the rectangle R. Let $\mathcal{Q}_\delta$ be the partition:

$$\mathcal{Q}_\delta := \{\{a_1, a_1 + \delta, b_1 - \delta, b_1\}, \ldots, \{a_m, a_m + \delta, b_m - \delta, b_m\}\}.$$

Let $C = \sup_{x \in R} |f(x)|$. Let

$$Q_\delta = [a_1 + \delta, b_1 - \delta] \times \cdots \times [a_m + \delta, b_m - \delta] \qquad (1.5.c)$$

be the big central rectangle of the partition. Let $\mathfrak{Q}_\delta := \mathcal{Q}_\delta - Q_\delta$ be the remaining rectangles of the partition. We have $\check{m}(f, R) \leq m(f, Q_\delta)$. We estimate:

$$\check{m}(f, R) \text{vol}(R) \leq m(f, Q_\delta) \text{vol}(R) = m(f, Q_\delta) \text{vol}(Q_\delta) + \sum_{Q \in \mathfrak{Q}_\delta} m(f, Q_\delta) \text{vol}(Q)$$

$$= m(f, Q_\delta) \text{vol}(Q_\delta) + \sum_{Q \in \mathfrak{Q}_\delta} m(f, Q) \text{vol}(Q) + \sum_{Q \in \mathfrak{Q}_\delta} \{m(f, Q_\delta) - m(f, Q)\} \text{vol}(Q)$$

$$\leq \check{m}(f, Q_\delta) \text{vol}(Q_\delta) + \sum_{Q \in \mathfrak{Q}_\delta} \check{m}(f, Q) \text{vol}(Q) + \sum_{Q \in \mathfrak{Q}_\delta} 2C \text{vol}(Q)$$

$$= \check{L}(f, \mathcal{Q}) + 2C \{\text{vol}(R) - \text{vol}(Q_\delta)\}.$$

If $\mathcal{P}$ is a more complicated partition, we perform a similar construction on each rectangle of $\mathcal{P}$ to construct a partition $\mathcal{Q}_\delta$ and for each $P \in \mathcal{P}$, we let $Q_\delta(P)$ be the corresponding rectangle defined in Equation (1.5.c). We have:

$$\check{L}(f, \mathcal{P}) \leq L(f, \mathcal{Q}_\delta) + 2C \sum_{P \in \mathcal{P}} \{\text{vol}(P) - \text{vol}(Q_\delta(P))\}.$$

The error can be made uniformly small as $\delta \to 0$. This shows $\check{L}(f, \mathcal{P}) \leq L(f, \mathcal{Q}) + \epsilon$; the proof of the remaining assertion is similar.   □

1.5.2 SETS OF CONTENT AND MEASURE ZERO. In view of Lemma 1.13, we may either take the inf and sup over all of a rectangle or over just the interior of a rectangle in future arguments involving integrals. We shall not belabor the point, but simply use Lemma 1.13 as necessary to simplify an argument. We now introduce just a bit of additional terminology from measure theory. Let S be a subset of $\mathbb{R}^m$.

1. We say S has *content zero* if given $\epsilon > 0$, there exist a finite number of rectangles $\{R_1, \ldots, R_n\}$ so $S \subset R_1 \cup \cdots \cup R_n$ and so $\text{vol}(R_1) + \cdots + \text{vol}(R_n) < \epsilon$.

2. We say S has *measure zero* if given $\epsilon > 0$, there exist a countable (or finite) number of rectangles $\{R_i\}$ so $S \subset R_1 \cup R_2 \cup \cdots$ and so $\text{vol}(R_1) + \text{vol}(R_2) + \cdots < \epsilon$.

By using an argument similar to that used to prove Lemma 1.13, we may work with open rectangles rather than with closed rectangles and define equivalently:

1. S has *content zero* if and only if given $\epsilon > 0$, there exist a finite number of rectangles $\{R_1, \ldots, R_n\}$ so $S \subset \text{int}(R_1) \cup \cdots \cup \text{int}(R_n)$ and so $\text{vol}(R_1) + \cdots + \text{vol}(R_n) < \epsilon$.

2. S has *measure zero* if and only if given $\epsilon > 0$, there exist a countable (or finite) number of rectangles $\{R_i\}$ so $S \subset \text{int}(R_1) \cup \text{int}(R_2) \cup \cdots$ and so $\text{vol}(R_1) + \text{vol}(R_2) + \cdots < \epsilon$.

Lemma 1.14

1. A finite union of sets of content zero has content zero.

2. A countable (or finite) union of sets of measure zero has measure zero.

3. Any countable (or finite) set has measure zero.

4. Any compact set of measure zero has content zero.

5. The closure of any set of content zero is a compact set of content zero.

Proof. Suppose sets $\{S_1, \ldots, S_n\}$ have content zero. Let $\epsilon > 0$ be given. Choose rectangles $R^j_{i_j}$ for $1 \leq i_j \leq v_j$ so

$$\sum_{i_j=1}^{v_j} \text{vol}(R^j_{i_j}) < 3^{-j}\epsilon \quad \text{and} \quad S_j \subset \bigcup_{i_j=1}^{v_j} R^j_{i_j} .$$

We show $S_1 \cup \cdots \cup S_n$ has content zero and establish Assertion 1 by observing:

$$\sum_{j=1}^{n} \sum_{i_j=1}^{v_j} \text{vol}(R^j_{i_j}) < \sum_{j=1}^{n} 3^{-j}\epsilon < \epsilon \quad \text{and} \quad \bigcup_{j=1}^{n} S_j \subset \bigcup_{j=1}^{n} \bigcup_{i_j=1}^{v_j} R^j_{i_j} .$$

Similarly, suppose we have a countable (or finite) collection of sets of measure zero. Let $\epsilon > 0$. Choose a countable (or finite) collection of rectangles $R_{i_j}^j$ so

$$\sum_{ij} \text{vol}(R_{i_j}^j) < 3^{-j}\epsilon \quad \text{and} \quad S_j \subset \bigcup_{ij} R_{i_j}^j .$$

Consider the countable (or finite) collection of rectangles $\{R_{i_j}^j\}$. We may show that $S_1 \cup \cdots$ has measure zero and establish Assertion 2 by observing:

$$\sum_j \sum_{ij} \text{vol}(R_{i_j}^j) < \sum_j 3^{-j}\epsilon < \epsilon \quad \text{and} \quad \bigcup_j S_j \subset \bigcup_{j,ij} R_{i_j}^j .$$

Assertion 3 is immediate since any singleton set has measure zero and since any finite collection is also a countable collection. Suppose C is a compact set of measure zero. Let $\epsilon > 0$ be given. As noted above, we can use open rectangles rather than closed rectangles to find a countable collection of rectangles R_i so

$$\sum_{i=1}^{\infty} \text{vol}(R_i) < \epsilon \quad \text{and} \quad C \subset \bigcup_{i=1}^{\infty} \text{int}(R_i) .$$

Since $\{\text{int}(R_i)\}$ is an open cover of the compact set C, there is a finite subcover

$$\{\text{int}(R_{i_1}) \cup \cdots \cup \text{int}(R_{i_k})\}$$

of C. This shows that C has content zero and establishes Assertion 4. Finally let S have content zero. Let $\epsilon > 0$ be given. We find a finite collection of *closed* rectangles so

$$\sum_{i=1}^{n} \text{vol}(R_i) < \epsilon \quad \text{and} \quad S \subset R_1 \cup \cdots \cup R_n .$$

Since $\text{closure}(S) \subset \text{closure}(R_1 \cup \cdots \cup R_n) = R_1 \cup \cdots \cup R_n$, we conclude the closure of S has content zero. A finite union of closed rectangles is compact. A closed subset of a compact set is compact and hence the closure of S is compact. $\square$

If f is a real-valued function on a rectangle $R \subset \mathbb{R}^m$, define the *graph* of f by:

$$G_f := \{(x, f(x)) \in \mathbb{R}^{m+1} : x \in R\}.$$

Lemma 1.15   If f is integrable, then G_f has content zero.

Proof. Let $\epsilon > 0$ be given. Since f is integrable, we may choose a partition $\mathcal{P} = \{P_i\}$ of the defining rectangle $R \subset \mathbb{R}^m$ so that $U(f, \mathcal{P}) - L(f, \mathcal{P}) < \varepsilon$. If $P \in \mathcal{P}$ is an m-dimensional rectangle, define an $m + 1$-dimensional rectangle $Q(P)$ by setting:

$$Q(P) := P \times [m(f, P), M(f, P)].$$

We have $G_f \subset \cup_{P \in \mathcal{P}} Q(P)$. This implies the following inequality from which the Lemma follows:

$$\sum_{P \in \mathcal{P}} \text{vol}(Q(P)) = \sum_{P \in \mathcal{P}} \{M(f, P) - m(f, P)\} \text{vol}(P) = U(f, \mathcal{P}) - L(f, \mathcal{P}) < \epsilon. \qquad \square$$

1.5.3 REMARK. The converse of Lemma 1.15 is false. Let f be the characteristic function of the rational numbers in $[0, 1]$; $f(x) = 1$ if $x \in [0, 1]$ is rational and $f(x) = 0$ if $x \in [0, 1]$ is not rational. By Theorem 1.17 below, f is not integrable since f is discontinuous at every point of $[0, 1]$. On the other hand $G_f \subset [0, 1] \times \{0\} \cup [0, 1] \times \{1\}$ has content zero in $\mathbb{R}^2$.

Let f be a bounded real-valued function defined on a set $S \subset \mathbb{R}^m$. If $0 < r < s$, then

$$\left\{ \sup_{x \in S \cap B_r(P)} f(x) - \inf_{x \in S \cap B_r(P)} f(x) \right\} \leq \left\{ \sup_{x \in S \cap B_s(P)} f(x) - \inf_{x \in S \cap B_s(P)} f(x) \right\}$$

so this difference is a monotonically decreasing non-negative function of r. We define the *oscillation of f* around a point P of S by setting:

$$o(f, P) := \lim_{r \to 0} \left\{ \sup_{x \in S \cap B_r(P)} f(x) - \inf_{x \in S \cap B_r(P)} f(x) \right\}. \qquad (1.5.d)$$

The following result follows immediately from the definition:

Lemma 1.16   Let f be a bounded real-valued function defined on a set $S \subset \mathbb{R}^m$.

1. f is continuous at $P \in S$ if and only if $o(f, P) = 0$.

2. Let $\epsilon > 0$. If S is closed, then $\{x \in S : o(f, x) \geq \epsilon\}$ is closed.

We can now give a useful characterization of what it means for a function to be Riemann integrable:

Theorem 1.17   *Let R be a rectangle which is contained in $\mathbb{R}^m$.*

1. *Let f be a bounded real-valued function on R. Let D be the set of all points of R where f is discontinuous. Then f is integrable on R if and only if D has measure zero.*

2. *If f and g are integrable on R, then fg is integrable on R.*

3. *If f is integrable on R and if g is continuous on the range of f, then $g \circ f$ is integrable on R.*

4. *If f is a non-decreasing function on $R = [a, b]$, then f is integrable.*

Proof. Before proving Assertion 1, we establish some additional notation. Since f is bounded, we may choose k so $|f| \leq k$ on the closed rectangle R. Let $o(f, x)$ be the oscillation of f about x. Let $D_n := \{x \in R : o(f, x) \geq \frac{1}{n}\}$. By Lemma 1.16, D_n is closed. Since $D_n \subset R$, D_n is bounded and hence compact. We have $D = D_1 \cup D_2 \cup \cdots$.

Suppose first that D has measure zero. Then each D_n has measure zero and hence, being compact, has content zero by Lemma 1.14. Let $\epsilon > 0$ be given. Choose $\epsilon_1 > 0$ and $n \in \mathbb{N}$ so

$$2k\epsilon_1 < \tfrac{1}{3}\epsilon \quad \text{and} \quad \tfrac{1}{n}\,\mathrm{vol}(R) < \tfrac{1}{3}\epsilon\,.$$

Since D_n has content zero, we may find a finite number of rectangles $\{R_1, \ldots, R_\mu\}$ so

$$D_n \subset R_1 \cup \cdots \cup R_\mu \quad \text{and} \quad \mathrm{vol}(R_1) + \cdots + \mathrm{vol}(R_\mu) < \epsilon_1\,.$$

By considering all the hyperplanes which form the sides of the rectangles R_i, we can construct a partition $\mathcal{P}$ of R and decompose $\mathcal{P} = \mathcal{P}_1 \cup \mathcal{P}_2$ as the disjoint union of two families of rectangles so

$$D_n \subset \bigcup_{P \in \mathcal{P}_1} P, \quad \sum_{P \in \mathcal{P}_1} \mathrm{vol}(P) < \epsilon_1, \quad \mathrm{int}(P) \cap D_n = \emptyset \quad \text{for} \quad P \in \mathcal{P}_2\,.$$

We then have

$$\sum_{P \in \mathcal{P}_1} \{M(f, P) - m(f, P)\}\,\mathrm{vol}(P) \leq 2k\epsilon_1 < \tfrac{1}{3}\epsilon\,. \tag{1.5.e}$$

For each rectangle $P \in \mathcal{P}_2$, use Lemma 1.13 to find a rectangle $Q(P) \subset \mathrm{int}(P)$ so that

$$\sum_{P \in \mathcal{P}_2} \{\mathrm{vol}(P) - \mathrm{vol}(Q(P))\} < \epsilon_1\,.$$

We use the same argument which was used to establish Equation (1.5.e) to show that:

$$\sum_{P \in \mathcal{P}_2} \{M(f, P) - m(f, P)\}\{\mathrm{vol}(P) - \mathrm{vol}(Q(P))\} \leq 2k\epsilon_1 < \tfrac{1}{3}\epsilon\,. \tag{1.5.f}$$

Since $Q(P) \subset \mathrm{int}(P)$ and $D_n \cap \mathrm{int}(P) = \emptyset$, for each $x \in Q(P)$, we can find $r(x) > 0$ so that:

$$|y - x| < r(x) \quad \Rightarrow \quad |f(x) - f(y)| < \tfrac{1}{n}\,.$$

Since $Q(P)$ is compact, by subdividing P if necessary to ensure that the diameter of $Q(P)$ is small and by using the Lebesgue covering lemma, we can construct a partition $\mathcal{Q}(P)$ of $Q(P)$ so that $|f(x) - f(y)| < \frac{1}{n}$ where x and y belong to any rectangle of $\mathcal{Q}(P)$. This shows

$$\sum_{P \in \mathcal{P}_2} \sum_{Q_1 \in \mathcal{Q}(P)} \{M(f, Q_1) - m(f, Q_1)\}\,\mathrm{vol}(Q_1) \leq \tfrac{1}{n}\,\mathrm{vol}(R) < \tfrac{1}{3}\epsilon\,. \tag{1.5.g}$$

We extend $\mathcal{Q}(P)$ to a refinement $\mathcal{Q}_1$ of $\mathcal{P}$ and use Equations (1.5.e), (1.5.f), and (1.5.g) to construct a partition with $U(f, \mathcal{Q}_1) - L(f, \mathcal{Q}_1) < \epsilon$. By Lemma 1.12, f is integrable.

Conversely, suppose that f is integrable. Let $\epsilon > 0$ be given. Fix n. Choose $\epsilon_1 > 0$ so that $n\epsilon_1 < \frac{1}{3}\epsilon$. Choose a partition $\mathcal{P}$ so that $U(f, \mathcal{P}) - L(f, \mathcal{P}) < \epsilon_1$. Let $\mathcal{P}_1$ be the set of rectangles $P \in \mathcal{P}$ so that $D_n \cap \operatorname{int}(P) \neq \emptyset$. The following picture is perhaps useful:

The rectangles of $\mathcal{P}_1$
where $\mathfrak{o}(f, x) \geq \frac{1}{n}$

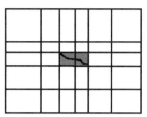

We then have

$$\frac{1}{n} \sum_{P \in \mathcal{P}_1} \operatorname{vol}(P) \leq \sum_{P \in \mathcal{P}_1} \{M(f, P) - m(f, P)\} \operatorname{vol}(P) \leq U(f, \mathcal{P}) - L(f, \mathcal{P}) < \epsilon_1 .$$

Consequently $\sum_{P \in \mathcal{P}_1} \operatorname{vol}(P) < \frac{1}{3}\epsilon$. There can, of course, be points of D_n which are not in the interior of any rectangle of $\mathcal{P}$. But such points necessarily lie on the boundary of some rectangle. The boundary of any rectangle necessarily has content zero. Thus D_n has content zero and hence measure zero. Since $D = \cup_n D_n$, D has measure zero by Lemma 1.14. This establishes Assertion 1; Assertion 2 and Assertion 3 then follow immediately.

Let f be a monotone non-decreasing real-valued function on an interval $[a, b]$ and let

$$\mathcal{P} = \{a = c_0 < c_1 < \cdots < c_m = b\}$$

be a partition of $[a, b]$. One may verify that one has $\sum_i \mathfrak{o}(f, c_i) \leq f(b) - f(a)$. From this it follows that the set of discontinuities of f is countable and thus of measure zero. This implies f is integrable which establishes Assertion 4.   □

We illustrate Theorem 1.17 by considering the function constructed in Example 1.5.1. This function is integrable and the set of discontinuities is the set of rational points in $[0, 1]$. Since this set is countable, the set of discontinuities has measure zero.

Let S be a subset of $\mathbb{R}^m$. We define the characteristic function of S by setting:

$$\chi_S(x) = \begin{cases} 1 & \text{if } x \in S \\ 0 & \text{if } x \notin S \end{cases} .$$

Lemma 1.18   Let R be a rectangle in $\mathbb{R}^m$ and let $S \subset R$.

1. The following conditions are equivalent and if either is satisfied, then S is said to be Jordan measurable and we set $\operatorname{vol}(S) = \int_R \chi_S$:

    (a) χ_S is integrable.

(b) The boundary of S has content zero.

2. S is Jordan measurable if and only if given any $\epsilon > 0$, there exists a partition $\mathcal{P}$ of R so that if $\mathcal{P}_1 := \{P \in \mathcal{P} : P \subset S\}$ and if $\mathcal{P}_2 := \{P \in \mathcal{P} : P \cap S \neq \emptyset\}$, then

$$\bigcup_{P \in \mathcal{P}_1} P \subset S \subset \bigcup_{P \in \mathcal{P}_2} P \quad \text{and} \quad \sum_{P \in \mathcal{P}_1} \mathrm{vol}(P) \leq \sum_{P \in \mathcal{P}_2} \mathrm{vol}(P) \leq \sum_{P \in \mathcal{P}_1} \mathrm{vol}(P) + \epsilon.$$

In this setting, we have $\displaystyle \sum_{P \in \mathcal{P}_1} \mathrm{vol}(P) \leq \mathrm{vol}(S) \leq \sum_{P \in \mathcal{P}_2} \mathrm{vol}(P) \leq \sum_{P \in \mathcal{P}_1} \mathrm{vol}(P) + \epsilon.$

3. If $\mathrm{vol}(R) > 0$, then R does not have measure zero.

Proof. Assertion 1 follows from Theorem 1.17 since the set of discontinuities of χ_S is exactly the boundary of S. Let $\mathcal{P}$ be a partition of R and let $P \in \mathcal{P}$. We have 3 cases:

1. If $P \cap S = \emptyset$, then $m(\chi_S, P) = M(\chi_S, P) = 0$.

2. If $P \cap S \neq \emptyset$ and $P \not\subset S$, then $m(\chi_S, P) = 0$ and $M(\chi_S, P) = 1$.

3. If $P \subset S$, then $m(\chi_S, P) = 1$ and $M(\chi_S, P) = 1$.

This shows that

$$\sum_{P \in \mathcal{P}_1} \mathrm{vol}(P) = L(\chi_S, \mathcal{P}) \leq U(\chi_S, \mathcal{P}) = \sum_{P \in \mathcal{P}_2} \mathrm{vol}(P).$$

If S is Jordan measurable, then χ_S is integrable. Thus given $\epsilon > 0$, there exists a partition $\mathcal{P}$ so $U(\chi_S, \mathcal{P}) - L(\chi_S, \mathcal{P}) < \epsilon$ and the desired inequality holds. Conversely, if given $\epsilon > 0$ we can find a suitable partition $\mathcal{P}$, then $U(\chi_S, \mathcal{P}) - L(\chi_S, \mathcal{P}) < \epsilon$ and χ_S is integrable. This proves Assertion 2.

Suppose that R is a rectangle in $\mathbb{R}^m$ with $\mathrm{Vol}(R) > 0$ but, to the contrary, that R has measure zero. Since R is compact, R has content zero. Let $R_1, \ldots, R_\ell$ be a cover of R by closed rectangles with $\mathrm{vol}(R_1) + \cdots + \mathrm{vol}(R_\ell) < \frac{1}{2}\mathrm{vol}(R)$. We have

$$\chi_R \leq \chi_{R_1} + \cdots + \chi_{R_\ell},$$
$$\mathrm{vol}(R) = \int \chi_R \leq \int \{\chi_{R_1} + \cdots + \chi_{R_\ell}\} = \int \chi_{R_1} + \cdots + \int \chi_{R_\ell}$$
$$= \mathrm{vol}(R_1) + \cdots + \mathrm{vol}(R_\ell) < \tfrac{1}{2}\mathrm{vol}(R).$$

This is not possible since $\mathrm{vol}(R) > 0$.   □

This is, of course, exactly the approach taken by Euclid to determine the area of a circle by examining inscribed and circumscribed polygons; if we set

$$S_1 := \cup_{P \in \mathcal{P}: P \subset S} P \quad \text{and} \quad S_2 := \cup_{P \in \mathcal{P}: P \cap S \neq \emptyset} P,$$

then S_1 and S_2 are unions of rectangles (and hence can be regarded as polygons) which satisfy the relations $S_1 \subset S \subset S_2$ and $\mathrm{vol}(S_1) \leq \mathrm{vol}(S) \leq \mathrm{vol}(S_2) \leq \mathrm{vol}(S_1) + \epsilon.$

Let S be a Jordan measurable subset of a rectangle R and let f be integrable on R. By Theorem 1.17, $\chi_S f$ is integrable and one sets

$$\int_S f := \int_R \chi_S f .$$

In particular, if we take $f = 1$, we see that $\int_S 1 = \mathrm{vol}(S)$. Clearly Jordan measurable sets are bounded. The following 1-dimensional example shows that not every bounded open set is Jordan measurable.

1.5.4 EXAMPLE. Let $\{r_n\}$ be an enumeration of the rationals in $(0, 1)$. Let

$$\mathcal{O}_n := B_{5^{-n}}(r_n) \cap (0, 1) \quad \text{and} \quad \mathcal{O} := \bigcup_{n=1}^{\infty} \mathcal{O}_n .$$

Note that $\mathcal{O}$ is an open subset of $(0, 1)$. We assume $\mathcal{O}$ is Jordan measurable and argue for a contradiction. Since $\mathcal{O}$ is Jordan measurable, the boundary $\mathrm{bd}(\mathcal{O})$ has measure zero. Let C be the complementary closed set of $[0, 1]$, $C = [0, 1] \cap \mathcal{O}^c$. Since $\mathcal{O}$ is a dense subset of $[0, 1]$,

$$\mathrm{bd}(\mathcal{O}) = \bar{\mathcal{O}} \cap \overline{\mathcal{O}^c} = [0, 1] \cap \mathcal{O}^c = C .$$

This implies C has measure zero. Then we can find a countable collection of rectangles R_i so $C \subset \mathrm{int}(R_1) \cup \cdots$ and so $\sum_i \mathrm{vol}(R_i) < \frac{1}{5}$. The collection $\{\mathcal{O}_n, \mathrm{int}(R_i)\}$ is an open cover of $[0, 1]$. Since $[0, 1]$ is compact, there is a finite subcover

$$[0, 1] \subset \mathcal{O}_1 \cup \cdots \cup \mathcal{O}_n \cup R_1 \cup \cdots \cup R_n .$$

Therefore $\chi_{[0,1]} \le \chi_{\mathcal{O}_1} + \cdots + \chi_{\mathcal{O}_n} + \chi_{R_1} + \cdots + \chi_{R_n}$. We derive the desired contradiction and show $\mathcal{O}$ is not Jordan measurable by integrating this inequality to see:

$$1 = \int \chi_{[0,1]} \le \int \chi_{\mathcal{O}_1} + \cdots + \int \chi_{\mathcal{O}_n} + \int \chi_{R_1} + \cdots + \int \chi_{R_n}$$

$$= \mathrm{vol}(\mathcal{O}_1) + \cdots + \mathrm{vol}(\mathcal{O}_n) + \mathrm{vol}(R_1) + \cdots + \mathrm{vol}(R_n) \le \sum_{i=1}^{n} \frac{2^i}{5^i} + \frac{1}{5} < 1 .$$

1.5.5 FUBINI'S THEOREM. Fubini's Theorem will permit us to evaluate multivariable integrals as iterated integrals. It is due to the Italian mathematician Guido Fubini.

Guido Fubini (1879–1943)

Theorem 1.19 (Fubini's Theorem).   *Let $A \subset \mathbb{R}^a$ and $B \subset \mathbb{R}^b$ be rectangles. Let f be a bounded real-valued integrable function on $A \times B$. Set $f_x(y) := f(x, y)$ for $x \in A$ and $y \in B$. Let $\mathcal{L}$ and $\mathcal{U}$ be the lower and upper integrals, i.e., $\mathcal{L}(x) := \underline{\int}_B f_x$ and $\mathcal{U}(x) := \overline{\int}_B f_x$. Then $\mathcal{L}$ and $\mathcal{U}$ are integrable and:*

$$\int_{A \times B} f = \int_A \mathcal{L} = \int_A \mathcal{U}.$$

Proof. Let $P \times Q$ be a partition of $A \times B$ where $\mathcal{P}$ is a partition of A and Q is a partition of B. Let $P \times Q \in \mathcal{P} \times Q$. For any $x \in P$, we have $m(f, P \times Q) \le m(f_x, Q)$. Consequently:

$$\sum_{Q \in Q} m(f, P \times Q) \operatorname{vol}(Q) \le \sum_{Q \in Q} m(f_x, Q) \operatorname{vol}(Q) = L(f_x, Q) \le \underline{\int}_B f_x = \mathcal{L}(x).$$

We take the inf over $x \in P$ to see

$$\sum_{Q \in Q} m(f, P \times Q) \operatorname{vol}(Q) \le m(\mathcal{L}, P).$$

Consequently, summing over $P \in \mathcal{P}$ yields:

$$
\begin{aligned}
L(f, \mathcal{P} \times Q) &= \sum_{P \in \mathcal{P}} \sum_{Q \in Q} m(f, P \times Q) \operatorname{vol}(Q) \operatorname{vol}(P) \\
&\le \sum_{P \in \mathcal{P}} m(\mathcal{L}, P) \operatorname{vol}(P) = L(\mathcal{L}, \mathcal{P}) \le \underline{\int}_A \mathcal{L}.
\end{aligned}
$$

Taking the sup over all partitions $\mathcal{P} \times Q$ then yields $\int_{A \times B} f \le \underline{\int}_A \mathcal{L}$. A similar argument yields $\overline{\int}_A \mathcal{U} \le \int_{A \times B} f$. Thus we have

$$\int_{A \times B} f \le \underline{\int}_A \mathcal{L} \le \overline{\int}_A \mathcal{L} \le \overline{\int}_A \mathcal{U} \le \int_{A \times B} f.$$

Since all these inequalities must have been equalities, $\mathcal{L}$ is integrable and $\int_{A \times B} f = \int_A \mathcal{L}$. A similar argument can be used to deal with $\mathcal{U}$. $\qquad\square$

It is necessary to introduce several caveats. These are best illustrated by a series of examples. If f is in fact continuous, then it is not necessary to go through the technical fuss of considering $\mathcal{L}$ and $\mathcal{U}$ as the double integral can simply be computed as the iterated integral. However in the general setting, it is necessary to be more careful. We have the following illustrative examples:

1. Let $A = B = [0, 1]$. Define

$$
f(x, y) := \left\{
\begin{array}{ll}
+1 & \text{if } x \text{ is rational and } y \le \frac{1}{2} \\
0 & \text{if } x \text{ is rational and } y > \frac{1}{2} \\
0 & \text{if } x \text{ is irrational and } y \le \frac{1}{2} \\
+1 & \text{if } x \text{ is irrational and } y > \frac{1}{2}
\end{array}
\right\}.
$$

Then f_x is integrable for any fixed x and $\int_0^1 f_x(y)dy = \frac{1}{2}$. We have that f is discontinuous on all of $[0, 1] \times [0, 1]$. By Lemma 1.18, $[0, 1] \times [0, 1]$ does not have measure zero and thus f is not integrable.

2. Let S be a countable dense subset of $R := [0, 1] \times [0, 1]$ so that any horizontal or vertical line contains at most one point of S. Then χ_S is discontinuous at every point of R and hence is not integrable. However, $\int_0^1 f_x(y)dy = 0$ for all x and $\int_0^1 f_y(x)dx = 0$ for all y so the iterated integrals exist.

3. Let $R := [0, 1] \times [0, 1]$. Let f be as defined in Example 1.5.1, i.e., if $\{r_n\}$ is an enumeration of the rationals in $[0, 1]$, then

$$f(x) := \left\{ \begin{array}{ll} \frac{1}{n} & \text{if } x = r_n \text{ for some n} \\ 0 & \text{if } x \neq r_n \text{ for all } n \end{array} \right\}.$$

Let

$$g(x, y) := \left\{ \begin{array}{ll} f(x) & \text{if } y \text{ is rational} \\ 0 & \text{if } y \text{ is irrational} \end{array} \right\}.$$

The argument given in Example 1.5.1 extends immediately to show g is integrable and $\int_R g = 0$. However, if x is rational, then $f(x) \neq 0$ so the integral $\int_{[0,1]} g_x(y)dy$ does not exist if x is rational and the iterated integral $\int_R g(x, y)dydx$ does not exist and it is necessary to deal with $\mathcal{U}(x) = f(x)$ and $\mathcal{L}(x) = 0$ in examining the iterated integrals. On the other hand $\int_{[0,1]} g_y(x)dx = 0$ for all y so the iterated integral $\int_R g(x, y)dxdy$ can be evaluated directly.

1.5.6 APPLICATIONS OF FUBINI'S THEOREM.

Theorem 1.20

1. **(Clairaut's Theorem [12])** *Let $\mathcal{O}$ be an open subset of $\mathbb{R}^2$. Let $f : \mathcal{O} \to \mathbb{R}$ be C^2. Then $\partial_x \partial_y f = \partial_y \partial_x f$.*

2. **(Cavalieri's Principle)** *Let A_i be Jordan measurable subsets of $\mathbb{R}^m$. For $\vec{x} \in \mathbb{R}^{m-1}$, consider the slice $A_i(c) := \{\vec{x} : (\vec{x}, c) \in A_i\}$. Suppose for any c that $A_i(c)$ are Jordan measurable subsets of $\mathbb{R}^{m-1}$ with $\text{vol}(A_1(c)) = \text{vol}(A_2(c))$. Then $\text{vol}(A_1) = \text{vol}(A_2)$.*

Cavalieri's principle was introduced by the Italian mathematician Bonaventura Francesco Cavalieri.

B. Cavalieri (1598–1647)

Proof. Suppose Assertion 1 fails; we argue for a contradiction. By interchanging the roles of x and y if necessary, we may assume that $\{\partial_x \partial_y f - \partial_y \partial_x f\}(P) > 0$ for some point $P \in \mathcal{O}$. Since the second partial derivatives are continuous, we can find a rectangle R so that $P \in R \subset \mathcal{O}$ and so that $\partial_x \partial_y f - \partial_y \partial_x f \geq \epsilon$ on R for some $\epsilon > 0$. Let $R = [a, b] \times [c, d]$. Consequently:

$$\int_R (\partial_x \partial_y f - \partial_y \partial_x f) \geq \epsilon \operatorname{vol}(R) > 0. \tag{1.5.h}$$

We use Theorem 1.19 and the Fundamental Theorem of Calculus to see:

$$
\begin{aligned}
\int_R \partial_x \partial_y f &= \int_c^d \int_a^b \partial_x \{\partial_y f\}(x, y) dx dy = \int_c^d \{\partial_y f(b, y) - \partial_y f(a, y)\} dy \\
&= f(b, d) + f(a, c) - f(a, d) - f(b, c), \\
\int_R \partial_y \partial_x f &= \int_a^b \int_c^d \partial_y \{\partial_x f\}(x, y) dy dx = \int_a^b \{\partial_x f(x, d) - \partial_x f(x, c)\} dy \\
&= f(b, d) + f(a, c) - f(a, d) - f(b, c).
\end{aligned}
$$

These two integrals are equal, which contradicts Equation (1.5.h). We have the following picture illustrating Assertion 2:

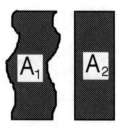

Introduce coordinates $(\vec{x}, t)$ on $\mathbb{R}^m$. We compute:

$$
\begin{aligned}
\operatorname{vol}(A_1) &= \int \int \chi(A_1)(\vec{x}, t) d\vec{x} dt = \int \operatorname{vol}(A_1(t)) dt \\
&= \int \operatorname{vol}(A_2(t)) dt = \int \int \chi(A_2)(\vec{x}, t) d\vec{x} dt = \operatorname{vol}(A_2).
\end{aligned}
$$
□

The assumption that the second partial derivatives are continuous in Assertion 1 is necessary as the following example shows. Let

$$f(x, y) := \begin{cases} \frac{xy(x^2-y^2)}{x^2+y^2} & \text{if } (x, y) \neq (0,0) \\ 0 & \text{if } (x, y) = (0,0) \end{cases}.$$

This is clearly C^∞ away from the origin. Suppose $x = 0$. Then $f(0, y) = 0$ so:

$$(\partial_x f)(0, y) = \lim_{\Delta x \to 0} \frac{f(0 + \Delta x, y) - f(0, y)}{\Delta x} = \lim_{\Delta x \to 0} \frac{f(\Delta x, y)}{\Delta x}$$

$$= \lim_{\Delta x \to 0} \frac{1}{\Delta x} \frac{\Delta x \cdot y(\Delta x^2 - y^2)}{\Delta x^2 + y^2} = -y$$

and thus $(\partial_y \partial_x f)(0, 0) = -1$. A similar calculation shows $(\partial_x \partial_y f)(0, 0) = +1$.

1.6   IMPROPER INTEGRALS

If ϕ is a real-valued function on $\mathbb{R}^m$, then the *support* of ϕ is the closure in $\mathbb{R}^m$ of the set of points where $\phi \neq 0$.

1.6.1 MESA FUNCTIONS AND PARTITIONS OF UNITY. Let $\mathcal{O} \subset \mathbb{R}^m$ be open.

1. A *compact exhaustion* of $\mathcal{O}$ is a countable collection of compact subsets $C_n \subset \mathcal{O}$ so that $\mathcal{O} = \cup_n C_n$ and $C_n \subset \text{int}(C_{n+1})$.

2. Let C be a compact subset of $\mathcal{O}$. A *mesa function for C which is supported in $\mathcal{O}$* is a smooth real-valued function ϕ on $\mathbb{R}^m$ so that $0 \leq \phi \leq 1$, so that $\phi \equiv 1$ on C, and so that support$\{\phi\} \subset \mathcal{O}$.

3. Let $\{\mathcal{O}_\alpha\}$ be open sets with $\mathcal{O}_\alpha \subset \mathcal{O}$ and $\cup_\alpha \mathcal{O}_\alpha = \mathcal{O}$. A locally finite *partition of unity* for $\mathcal{O}$ which is subordinate to the cover $\{\mathcal{O}_\alpha\}$ is a countable collection of smooth real-valued functions ϕ_n on $\mathbb{R}^m$ which take values in $[0, 1]$ so:

   (a) For each point $x \in \mathcal{O}$, there is a neighborhood $\mathcal{U}(x)$ of x so that only a finite number of the functions ϕ_n are non-zero on $\mathcal{U}(x)$ (*locally finite*).

   (b) If $x \in \mathcal{O}$, then $\sum_n \phi_n(x) = 1$ (*partition of unity*).

   (c) For each n, there exists $\alpha(n)$ so that support$\{\phi_n\} \subset \mathcal{O}_{\alpha(n)}$ (*subordinate to the cover*).

Lemma 1.21   Let $C \subset \mathcal{O}$, where C is compact and $\mathcal{O}$ is a non-empty open subset of $\mathbb{R}^m$, and let $\{\mathcal{O}_\alpha\}$ be a cover of $\mathcal{O}$.

1. Let $\epsilon > 0$ be given. There exist smooth non-negative functions on $\mathbb{R}$ so that:

   (a) $f_0(x) = 0$ for $x \leq 0$ and $f_0(x) > 0$ for $x > 0$.

   (b) $f_{1,\epsilon}(x) = 0$ for $x \leq 0$ or $x \geq \epsilon$ and $f_{1,\epsilon}(x) > 0$ for $0 < x < \epsilon$.

   (c) $f_{2,\epsilon}(x) = 0$ for $x \leq 0$, $0 \leq f_{2,\epsilon}(x) \leq 1$ for $0 \leq x \leq \epsilon$, and $f_{2,\epsilon}(x) = 1$ for $x \geq \epsilon$.

   (d) $f_{3,\epsilon}(x) = 1$ for $x \leq \epsilon$, $f_{3,\epsilon}(x) = 0$ for $x \geq 2\epsilon$, and $0 \leq f_{3,\epsilon}(x) \leq 1$ for $\epsilon \leq x \leq 2\epsilon$.

2. There exists a compact exhaustion of $\mathcal{O}$ by Jordan measurable sets.

3. There exists a mesa function ϕ for C supported in $\mathcal{O}$.

4. There exists a locally finite partition of unity for $\mathcal{O}$ subordinate to the cover $\{\mathcal{O}_\alpha\}$.

Proof. To focus our discussion in the proof of Assertion 1, we give graphs of the 4 functions:

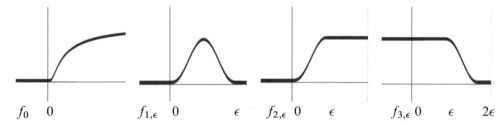

$f_0 \quad 0$ 　　　　　 $f_{1,\epsilon} \quad 0 \qquad \epsilon$ 　　　 $f_{2,\epsilon} \ 0 \qquad \epsilon$ 　　　 $f_{3,\epsilon} \ 0 \qquad \epsilon \qquad 2\epsilon$

Let $\epsilon > 0$ be given. Let

$$f_0(t) := \left\{ \begin{array}{ll} 0 & \text{if } t \leq 0 \\ e^{-t^{-1}} & \text{if } t > 0 \end{array} \right\}.$$

Clearly f_0 has the properties desired in Assertion 1-a for $t \neq 0$ so it suffices to show f_0 is smooth at 0. Since $e^t \geq \frac{t^n}{n!}$ for $t > 0$, we have $e^{t^{-1}} \geq \frac{1}{n!} t^{-n}$ and consequently

$$e^{-t^{-1}} \leq n! t^n \quad \text{for } t > 0 \text{ and for any } n \in \mathbb{N}. \tag{1.6.a}$$

We assume inductively that there is some polynomial $p_k(\cdot)$ so that

$$f_0^{(k)}(t) = \left\{ \begin{array}{ll} 0 & \text{if } t \leq 0 \\ p_k(t^{-1}) e^{-t^{-1}} & \text{if } t > 0 \end{array} \right\}. \tag{1.6.b}$$

We set $p_0 = 1$ and $k = 0$ to begin the induction. We use Equation (1.6.a) to see

$$\lim_{t \to 0} p_k(t^{-1}) e^{-t^{-1}} = 0$$

and thus $f_0^{(k)}$ is continuous at $t = 0$. Differentiating Equation (1.6.b) yields

$$f_0^{(k+1)}(t) = \left\{ \begin{array}{ll} 0 & \text{if } t < 0 \\ p_{k+1}(t^{-1})e^{-t^{-1}} & \text{if } t > 0 \end{array} \right\}$$

where $p_{k+1}(t^{-1}) = -t^{-2}\{p_k'(t^{-1}) - p_k(t^{-1})\}$. The left hand derivative of $f_0^{(k)}$ vanishes at 0. We complete the inductive step and establish Assertion 1-a by using Equation (1.6.a) to compute the right hand derivative:

$$\lim_{t\to 0^+} \frac{f_0^{(k)}(t) - f_0^{(k)}(0)}{t} = \lim_{t\to 0} t^{-1} p_k(t^{-1})e^{-t^{-1}} = 0.$$

The function $f_{1,\epsilon}(t) := f_0(t) f_0(\epsilon - t)$ has the properties of Assertion 1-b. The function

$$f_{2,\epsilon}(t) = \left\{ \int_{-\infty}^{\infty} f_{1,\epsilon}(s)ds \right\}^{-1} \int_{-\infty}^{t} f_{1,\epsilon}(s)ds \qquad (1.6.c)$$

has the properties of Assertion 1-c and the function $f_{3,\epsilon}(t) := 1 - f_{2,\epsilon}(t - 2\epsilon)$ has the properties of Assertion 1-d. This establishes Assertion 1.

If $\mathcal{O} = \mathbb{R}^m$, we set $C_k = B_k(0)$. Otherwise, we set:

$$C_k := \{x \in \mathcal{O} : \|x\| \leq k \text{ and } \text{dist}(x, \mathcal{O}^c) \geq \tfrac{1}{k}\}.$$

The collection $\{C_k\}$ forms a compact exhaustion of $\mathcal{O}$; however the sets C_k need not be Jordan measurable. As C_1 is compact, we can find a finite number of rectangles R_1^i contained in $\mathcal{O}$ so $C_1 \subset \cup_i \text{int}(R_1^i)$. Set $D_1 := \cup_i R_1^i$. Because $\text{bd}(D_1) \subset \cup_i \text{bd}(R_1^i)$, D_1 is Jordan measurable. We have $C_1 \subset \text{int}(D_1)$. Apply the same argument to the compact set $C_2 \cup D_1$ to find a number of rectangles R_2^i contained in $\mathcal{O}$ so $D_1 \cup C_2 \subset \cup_i \text{int}(R_2^i)$. Set $D_2 := \cup_i R_2^i$. We continue in this fashion to construct a sequence of compact Jordan measurable sets D_k contained in $\mathcal{O}$ so that $D_k \cup C_k \subset \text{int}(D_{k+1}) \subset D_{k+1}$. Since $\cup_k C_k = \mathcal{O}$, we have $\cup_k D_k = \mathcal{O}$. This establishes Assertion 2.

Let C be a compact subset of $\mathcal{O}$. Find a finite number of points x_i and associated radii $\epsilon_i > 0$ for $1 \leq i \leq \ell$ so $B_{3\epsilon_i}(x_i) \subset \mathcal{O}$ and so $C \subset \cup_i \text{int}(B_{\epsilon_i}(x))$. Set

$$\psi(x) := \sum_{i=1}^{\ell} f_{3,\epsilon_i^2}(\|x - x_i\|^2);$$

ψ is smooth and non-negative, ψ has support in $\mathcal{O}$ and ψ is at least 1 on C. We cut off ψ to have maximum value 1 and establish Assertion 3 by defining $\phi(x) := f_{2,1}(\psi(x))$.

Let $\{C_n\}$ be a compact exhaustion of $\mathcal{O}$. Let $R_n := C_n - \text{int}(C_{n-1})$; the "ring" R_n is a compact subset of $\mathcal{O}$. Fix n. We cover R_n by a finite number of balls $B_{r_n^i}(x_n^i)$ so

$$B_{3r_n^i}(x_n^i) \subset R_{n-1} \cup R_n \cup R_{n+1} \quad \text{and} \quad B_{3r_n^i}(x_n^i) \subset \mathcal{O}_{\alpha_n^i} \quad \text{for some} \quad \alpha_n^i.$$

The following picture will focus our discussion in the proof of Assertion 4 where for the purposes of illustration we suppose $C_n = \{x : \|x\| \leq n\}$ and $R_n = \{x : n-1 \leq \|x\| \leq n\}$; we show two small balls $B_{3r_n^i}(x_n^i)$ contained in the larger ring $R_{n-1} \cup R_n \cup R_{n+1}$:

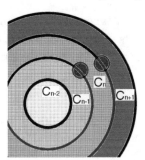

For each (i, n), let ψ_n^i be a mesa function for the closure of $B_{r_n^i}(x_n^i)$ which are supported in $B_{2r_n^i}(x_n^i)$. If $|n - k| \geq 5$, then ψ_k^i vanishes on $R_{n-1} \cup R_n \cup R_{n+1}$ which is a neighborhood of R_n. Thus the collection $\{\psi_n^i\}$ is a locally finite family and $\Psi(x) = \sum_{n,i} \psi_n^i$ is smooth on $\mathcal{O}$ (although it need not be defined on $\mathcal{O}^c$). Since the $\{B_{r_n^i}(x_n^i)\}$ covers R_n and since every x belongs to R_n for some n, we have $\Psi(x) \geq 1$ on $\mathcal{O}$. Set $\phi_n^i := \psi_n^i \Psi^{-1}$. Since ψ_n^i has support in $B_{2r_n^i}(x_n^i) \subset \mathcal{O}$, we can extend ϕ_n^i to $B_{3r_n^i}(x_n^i)^c$ to be identically zero to obtain a function which is smooth on all of $\mathbb{R}^m$. As each $B_{2r_n^i}(x_n^i)$ is contained in $\mathcal{O}_{\alpha_n^i}$, the collection is subordinate to the given cover. Assertion 4 now follows as $\sum_{n,i} \phi_n^i = 1$ on $\mathcal{O}$. $\qquad\square$

Let f be a real-valued function on an open subset $\mathcal{O}$ of $\mathbb{R}^m$. We say that f is *locally bounded* if given any compact set $C \subset \mathcal{O}$, there exists a constant κ_C so that $|f(x)| \leq \kappa_C$ for all $x \in C$. We say that f is *continuous almost everywhere* if the set of discontinuities of f has measure zero.

Lemma 1.22   Let f be a non-negative real-valued function on an open set $\mathcal{O}$ which is locally bounded and continuous almost everywhere. Let $\{\phi_n\}$ be a partition of unity subordinate to the cover of $\mathcal{O}$ by the interior of all the closed rectangles contained in $\mathcal{O}$, let $\{\psi_k\}$ be a collection of mesa functions for a compact exhaustion $\{A_k\}$ of $\mathcal{O}$, and let B_ℓ be a compact exhaustion of $\mathcal{O}$ by Jordan measurable sets. Then:

$$\sum_{n=1}^{\infty} \int \phi_n f = \lim_{k \to \infty} \int \psi_k f = \lim_{\ell \to \infty} \int_{B_\ell} f \,.$$

Proof. By definition $\phi_n f$ has support in the interior of a rectangle R_n which is contained in $\mathcal{O}$. Consequently, $\phi_n f$ is bounded on R_n, and $\phi_n f$ is continuous almost everywhere on R_n. Thus $\int \phi_n f$ is well-defined and non-negative. Consequently, the infinite sum in question is well-defined. It can, of course, be infinite. Next note that $\psi_k f$ has support in a compact set and hence $\psi_k f$ is bounded and continuous almost everywhere. Thus $\int \psi_k f$ is well-defined. Furthermore, f is non-negative and $\psi_k \leq \psi_{k+1}$ implies $\int \psi_k f \leq \int \psi_{k+1} f$ and thus $\lim_{k \to \infty} \int \psi_k f$ is well-defined; this limit can, of course, be infinite. Finally, note that B_ℓ is Jordan measurable, that f is bounded on B_ℓ, and that f is continuous almost everywhere on B_ℓ so $\int_{B_\ell} f$ is well-defined. Since $B_\ell \subset B_{\ell+1}$ and f is non-negative, $\int_{B_\ell} f \le \int_{B_{\ell+1}} f$ and thus $\lim_{\ell\to\infty} \int_{B_\ell} f$ is well-defined; the limit can, of course, be infinite. We establish the proof in three steps:

Step 1. Fix N. Then the support of $\phi_1 + \cdots + \phi_N$ is contained in a finite union of rectangles and hence is compact. Since $\mathcal{O} = \cup \, \text{int}(A_k)$, there exists K so the support of $\phi_1 + \cdots + \phi_N$ (which is a compact set) is contained in A_K (a compact set covered by an increasing union of open sets is eventually contained in a single set). Thus $\phi_1 + \cdots + \phi_N \le \psi_K$ so:

$$\sum_{n=1}^{N} \int \phi_n f = \int \left(\sum_{n=1}^{N} \phi_n \right) f \le \int \psi_K f \le \lim_{k\to\infty} \int \psi_k f \, .$$

As N was arbitrary, we may take the limit as $N \to \infty$ to show:

$$\sum_{n=1}^{\infty} \int \phi_n f \le \lim_{k\to\infty} \int \psi_k f \, . \tag{1.6.d}$$

Step 2. Fix k. As A_{k+1} is compact, the previous argument shows there exists L so $A_{k+1} \subset B_L$. Since ψ_k has support in A_{k+1}, $\psi_k \le \chi_{A_{k+1}} \le \chi_{B_L}$ and thus

$$\int \psi_k f \le \int \chi_{B_L} f = \int_{B_L} f \le \lim_{\ell\to\infty} \int_{B_\ell} f \, .$$

As k was arbitrary, we may take the limit as $k \to \infty$ to see

$$\lim_{k\to\infty} \int \psi_k f \le \lim_{\ell\to\infty} \int_{B_\ell} f \, . \tag{1.6.e}$$

Step 3. Fix ℓ. Since B_ℓ is a compact set and the $\{\phi_n\}$ are locally finite, only a finite number of the ϕ_n are non-zero on B_ℓ. In particular, we can choose N so $\phi_n = 0$ for $n > N$. Since $\{\phi_n\}$ is a partition of unity, $\phi_1 + \cdots + \phi_N = 1$ on B_ℓ. Thus $\chi_{B_\ell} \le \phi_1 + \cdots + \phi_N$ so:

$$\int_{B_\ell} f = \int \chi_{B_\ell} f \le \int \left(\sum_{n=1}^{N} \phi_n \right) f = \sum_{n=1}^{N} \int \phi_n f \le \sum_{n=1}^{\infty} \int \phi_n f \, .$$

As ℓ was arbitrary, we may take the limit as $\ell \to \infty$ to see that:

$$\lim_{\ell\to\infty} \int_{B_\ell} f \le \sum_{n=1}^{\infty} \int \phi_n f \, . \tag{1.6.f}$$

The Lemma now follows from Equations (1.6.d), (1.6.e), and (1.6.f). □

1.6.2 INTEGRABILITY IN THE EXTENDED SENSE. Let f be continuous except on a set of measure zero and locally bounded. If any (and hence all) of the 3 terms in Lemma 1.22 are finite, we say that f is *integrable in the extended sense* and we **define**

$$\int_{\mathcal{O}}^e f := \sum_{n=1}^\infty \int \phi_n f = \lim_{k\to\infty} \int \psi_k f = \lim_{\ell\to\infty} \int_{B_\ell} f .$$

The value is independent of partition of unity $\{\phi_n\}$, of the mesa functions $\{\psi_n\}$, and of the compact exhaustion by Jordan measurable sets $\{B_\ell\}$.

We have had to deal with a fair amount of technical fuss. The difficulty is, of course, that for the Riemann integral one does not have in general that the integral of an infinite sum is the infinite sum of the integrals nor is it true in general that the limit of the integrals is the integral of the limit; such theorems are, of course, the reason the Lebesgue integral was developed. We now show that the integral in the extended sense agrees with the ordinary integral if $\mathcal{O}$ is Jordan measurable and we shall subsequently replace $\int^e$ simply by $\int$.

Lemma 1.23   Let f be a non-negative real-valued function which is bounded and continuous almost everywhere on a Jordan measurable open set $\mathcal{O}$. Then $\int_{\mathcal{O}} f = \int_{\mathcal{O}}^e f$.

Proof. Choose a rectangle R so $\mathcal{O} \subset R$. Choose κ so $|f| \le \kappa$. Let $\epsilon > 0$ be given. Choose $\epsilon_1 > 0$ so $\kappa \epsilon_1 < \epsilon$. By Lemma 1.18, there exists a partition $\mathcal{P}$ of R so if

$$K := \cup_{P\in\mathcal{P}: P \subset \mathcal{O}} P ,$$

then K is a compact Jordan measurable subset of $\mathcal{O}$ with $\text{vol}(\mathcal{O} - K) < \epsilon_1$. Since we can embed K as the first term of a compact exhaustion of $\mathcal{O}$ by Jordan measurable sets,

$$\int_K f \le \int_{\mathcal{O}}^e f .$$

This shows that:

$$\int_{\mathcal{O}} f = \int_K f + \int_{\mathcal{O}-K} f \le \int_{\mathcal{O}}^e f + \kappa \, \text{vol}(\mathcal{O} - K) \le \int_{\mathcal{O}}^e f + \epsilon .$$

Since ϵ is arbitrary, we have

$$\int_{\mathcal{O}} f \le \int_{\mathcal{O}}^e f .$$

On the other hand if $\{C_\ell\}$ is a compact exhaustion of $\mathcal{O}$ by Jordan measurable sets, then $C_\ell \subset \mathcal{O}$ implies $\chi_{C_\ell} \le \chi_{\mathcal{O}}$ and hence $\int_{C_\ell} f \le \int_{\mathcal{O}} f$. Since this holds for any ℓ, we may take the sup over ℓ to obtain the reverse inequality and complete the proof; $\int_{\mathcal{O}}^e f \le \int_{\mathcal{O}} f$. $\qquad\square$

1.6.3 ABSOLUTELY INTEGRABLE FUNCTIONS. Let f be continuous almost everywhere and locally bounded on an open set $\mathcal{O}$. We say that f is *absolutely integrable* if $\int_{\mathcal{O}} |f| < \infty$. We set $f_+ := \frac{1}{2}(f + |f|)$ and $f_- := \frac{1}{2}(-f + |f|)$. By Theorem 1.17, the functions $f_{\pm}$ are continuous almost everywhere; they are clearly locally bounded as well.

$$f_+(x) = \left\{ \begin{array}{ll} f(x) & \text{if } f(x) \geq 0 \\ 0 & \text{if } f(x) \leq 0 \end{array} \right\} \quad \text{and} \quad f_-(x) = \left\{ \begin{array}{ll} 0 & \text{if } f(x) \geq 0 \\ -f(x) & \text{if } f(x) \leq 0 \end{array} \right\}.$$

We then have $f_{\pm} \leq |f|$ and hence the $f_{\pm}$ are integrable in the extended sense. Define:

$$\int_{\mathcal{O}} f := \int_{\mathcal{O}} f_+ - \int_{\mathcal{O}} f_-.$$

Lemma 1.24   Let f be absolutely integrable in the extended sense on $\mathcal{O}$. Adopt the notation of Lemma 1.22. Then

$$\int_{\mathcal{O}} f = \sum_{n=1}^{\infty} \int \phi_n f = \lim_{k \to \infty} \int \psi_k f = \lim_{\ell \to \infty} \int_{B_\ell} f.$$

We remark that then all the usual properties of the integral continue to hold. We shall not belabor the point. The advantage of the extended integral is that we can assign a volume (possibly infinite of course) to an open subset of $\mathbb{R}^m$. For example, let $\mathcal{O} \subset [0, 1]$ be the open subset of Example 1.5.4 that is not Jordan measurable. Since $\mathcal{O}$ is a bounded open set, we may define $\text{vol}(\mathcal{O}) := \int_{\mathcal{O}} 1 < \infty$. The following are other examples which further illustrate improper integrals phenomena.

1. Let $f(x) = \frac{1}{\sqrt{x}}$ on $(0, 1)$; f is continuous, non-negative, and locally bounded. We consider the compact exhaustion defined by $B_n := [\frac{1}{n}, 1 - \frac{1}{n}]$ for $n \geq 2$. Then

$$\int_0^1 f(x)dx = \lim_{n \to \infty} \int_{\frac{1}{n}}^{1-\frac{1}{n}} x^{-\frac{1}{2}} dx = \lim_{n \to \infty} 2x^{\frac{1}{2}} \Big|_{\frac{1}{n}}^{1-\frac{1}{n}} = \lim_{n \to \infty} 2\{(1 - \tfrac{1}{n})^{\frac{1}{2}} - (\tfrac{1}{n})^{\frac{1}{2}}\} = 2.$$

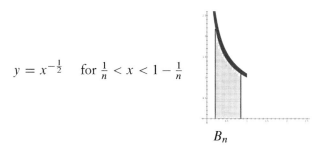

$$y = x^{-\frac{1}{2}} \quad \text{for } \tfrac{1}{n} < x < 1 - \tfrac{1}{n}$$

B_n

2. Let $f(x) = x^{-2}$ on $(1, \infty)$. Then this is continuous, non-negative, and locally bounded. Let $B_n := [1 + \frac{1}{n}, n]$ for $n \geq 2$. Then

$$\int_1^\infty x^{-2}\,dx = \lim_{n\to\infty} \int_{1+\frac{1}{n}}^n x^{-2}\,dx = \lim_{n\to\infty} -x^{-1}\Big|_{1+\frac{1}{n}}^n = \lim_{n\to\infty}\{(1+\tfrac{1}{n})^{-1} - \tfrac{1}{n}\} = 1 .$$

$$y = x^{-2} \quad \text{for } 1+\tfrac{1}{n} < x < n$$

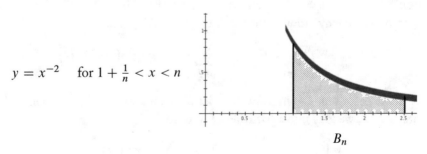

B_n

1.7   THE CHANGE OF VARIABLE THEOREM

Let $\mathcal{P}$ be a partition of a rectangle $R \subset \mathbb{R}^m$. If S is a Jordan measurable subset of R, define:

$$\mathcal{P}^1(S) := \{P \in \mathcal{P} : P \subset S\}, \quad \mathcal{P}^3(S) := \{P \in \mathcal{P} : P \subset S^c\},$$
$$\mathcal{P}^2(S) := \{P \in \mathcal{P} : P \cap S \neq \emptyset \text{ and } P \cap S^c \neq \emptyset\} . \tag{1.7.a}$$

We then have

$$\bigcup_{P \in \mathcal{P}^1(S)} P \subset S \subset \left(\bigcup_{P \in \mathcal{P}^1(S)} P \cup \bigcup_{P \in \mathcal{P}^2(S)} P \right), \quad \text{and}$$
$$\sum_{P \in \mathcal{P}^1(S)} \mathrm{vol}(P) \leq \mathrm{vol}(S) \leq \sum_{P \in \mathcal{P}^1(S)} \mathrm{vol}(P) + \sum_{P \in \mathcal{P}^2(S)} \mathrm{vol}(P) . \tag{1.7.b}$$

Assertion 2 of Lemma 1.18 shows that we can always choose $\mathcal{P}$ so Equation (1.7.b) is a good approximation, i.e., so $\sum_{P \in \mathcal{P}^2(S)} \mathrm{vol}(P)$ is as small as we like. We say that C is a *cube* if C is a rectangle in $\mathbb{R}^m$ so that all the sides have the same length $c > 0$. We extend Lemma 1.18 to use cubes rather than rectangles as follows:

Lemma 1.25   Let C be a cube of side c in $\mathbb{R}^m$. Let $\mathcal{P}_n$ be the partition of C into n^m cubes of side $\frac{c}{n}$. Let S be a Jordan measurable subset of C and let $\epsilon > 0$ be given. There exists $N = N(S,\epsilon)$ so that if $n \geq N$, then:

1. $\sum_{P \in \mathcal{P}_n^2(S)} \mathrm{vol}(P) < \epsilon$.

2. $\mathrm{vol}(S) - \sum_{P \in \mathcal{P}_n : P \subset S} \mathrm{vol}(P) < \epsilon$ and $\sum_{P \in \mathcal{P}_n : P \cap S \neq \emptyset} \mathrm{vol}(P) - \mathrm{vol}(S) < \epsilon$.

3. If S has content zero, then $\sum_{P \in \mathcal{P}_n : P \cap S \neq \emptyset} \mathrm{vol}(P) < \epsilon$.

Proof. Suppose first that S is a rectangle which is contained in C. If $m = 1$, then there are at most four intervals in $\mathcal{P}_n$ which touch the boundary of the interval S. More generally, there are at most $4mn^{m-1}$ rectangles which touch the boundary of S. Since $P \in \mathcal{P}_n^2(S)$ implies P touches the boundary of S, we may estimate (using Equation (1.7.b)) that:

$$\text{vol}(S) - \sum_{P \in \mathcal{P}_n^1(S)} \text{vol}(P) \leq \sum_{P \in \mathcal{P}_n^2(S)} \text{vol}(P) \leq 4mn^{m-1}(c/n)^m . \tag{1.7.c}$$

This tends to zero as n tends to infinity. Next let S be an arbitrary Jordan measurable subset of C. By Lemma 1.18, we may find a partition $\mathcal{Q}$ of C so

$$\text{vol}(S) - \sum_{Q \in \mathcal{Q}^1(S)} \text{vol}(Q) \leq \sum_{Q \in \mathcal{Q}^2(S)} \text{vol}(Q) \leq \tfrac{1}{4}\epsilon . \tag{1.7.d}$$

Applying the estimate of Equation (1.7.c) to the rectangles in $\mathcal{Q}^1(S)$ shows there is N_1 so

$$\sum_{Q \in \mathcal{Q}^1(S)} \text{vol}(Q) - \sum_{P \in \mathcal{P}_n : \exists Q \in \mathcal{Q} \text{ so } P \subset Q \subset S} \text{vol}(P) < \tfrac{1}{4}\epsilon \quad \text{for } n \geq N_1 . \tag{1.7.e}$$

Since $P \in \mathcal{P}_n^1(Q)$ for some $Q \in \mathcal{Q}^1(S)$ implies $P \in \mathcal{P}_n^1(S)$, we may use Equation (1.7.d) and Equation (1.7.e) to see that:

$$\text{vol}(S) - \sum_{P \in \mathcal{P}_n^1(S)} \text{vol}(P) \leq \text{vol}(S) - \sum_{P \in \mathcal{P}_n : \exists Q \in \mathcal{Q}^1(S) \text{ so } P \subset Q} \text{vol}(P) < \tfrac{1}{4}\epsilon + \tfrac{1}{4}\epsilon = \tfrac{1}{2}\epsilon .$$

Similarly we have

$$\text{vol}(C - S) - \sum_{P \in \mathcal{P}_n^3(S)} \text{vol}(P) < \tfrac{1}{2}\epsilon \quad \text{for} \quad n \geq N_2 .$$

Since $\text{vol}(S) + \text{vol}(C - S) = \text{vol}(C)$, we may establish Assertion 1 by computing:

$$\sum_{P \in \mathcal{P}_n^2(S)} \text{vol}(P) = \text{vol}(C) - \sum_{P \in \mathcal{P}_n^1(S)} \text{vol}(P) - \sum_{P \in \mathcal{P}_n^3(S)} \text{vol}(P)$$

$$= \text{vol}(S) - \sum_{P \in \mathcal{P}_n^1(S)} \text{vol}(P) + \text{vol}(C - S) - \sum_{P \in \mathcal{P}_n^3(S)} \text{vol}(P) < \epsilon .$$

Assertion 2 follows from Assertion 1 and from Equation (1.7.b). Assertion 3 follows from Assertion 2 since S has content zero implies $\text{vol}(S) = 0$ so $\mathcal{P}_n^1(S)$ is empty. □

If C is a cube $[a_1, a_1 + c] \times \cdots \times [a_m, a_m + c]$ for $c > 0$, then the *center of the cube* is the point $x_C := (a_1 + \tfrac{c}{2}, \dots, a_m + \tfrac{c}{2})$. Since the diameter of C is $\sqrt{m}c$, C is contained in the closed ball of radius $\tfrac{\sqrt{m}c}{2}$ about x_C. Correspondingly, the closed ball of radius $\tfrac{\sqrt{m}c}{2}$ about x_C is contained in a cube of side $\sqrt{m}c$ with center x_C. This geometric observation is at the heart of what follows.

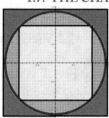

Lemma 1.26   Let $\mathcal{O}$ be an open subset of $\mathbb{R}^m$. Let $\Psi : \mathcal{O} \to \mathbb{R}^m$ be continuously differentiable. Let $C \subset \mathcal{O}$ be a cube of side c with center x_C. Let $\kappa := \sup_{P \in C} \|\Psi'(P)\|$ and let $S \subset C$ be Jordan measurable.

1. $\Psi(C)$ is contained in the cube of side $\sqrt{m}\kappa c$ and center $\Psi(x_C)$.

2. If S has content zero, then $\Psi(S)$ has content zero.

3. Assume $\det(\Psi') \neq 0$ and that Ψ is injective on $\mathcal{O}$. Then:

    (a) $\Psi(S)$ is Jordan measurable.

    (b) If Ψ is linear, then $\operatorname{vol}(\Psi(S)) = |\det(\Psi')| \cdot \operatorname{vol}(S)$.

    (c) If there exists $T_0 \in \operatorname{GL}(\mathbb{R}^m)$ so that $\|\Psi' - T_0\| < \epsilon$ on C, then
    $$\operatorname{vol}(\Psi(S)) \leq |\det(T_0)|(1 + \|T_0^{-1}\|\sqrt{m}\epsilon)^m \operatorname{vol}(S).$$

Proof. Let x_C be the center of the cube C. If $x_1 \in C$, then $\|x_1 - x_C\| \leq \frac{\sqrt{m}c}{2}$. Thus by Lemma 1.7, $\|\Psi(x_1) - \Psi(x_C)\| \leq \frac{\sqrt{m}\kappa c}{2}$. Consequently, $\Psi(C)$ is contained in the cube $\tilde{C}$ with side $\sqrt{m}\kappa c$ and center $\Psi(x_C)$. This proves Assertion 1.

Suppose that S has content zero. Let $\epsilon > 0$ be given. Choose ϵ_1 so $(\kappa\sqrt{m})^m \epsilon_1 < \epsilon$. Apply Lemma 1.25 to cover S by a collection of cubes C_i of side c/n so that

$$\sum_i \operatorname{vol}(C_i) < \epsilon_1.$$

Because $\Psi(S)$ is contained in the union of the sets $\Psi(C_i)$ and because the sets $\Psi(C_i)$ are contained in the cubes $\tilde{C}_i$ of side $\sqrt{m}\kappa c/n$, the cubes $\tilde{C}_i$ cover $\Psi(S)$. We observe that

$$\operatorname{vol}(\tilde{C}_i) = (\kappa\sqrt{m}c/n)^m = (\kappa\sqrt{m})^m \operatorname{vol}(C_i).$$

Assertion 2 follows from the estimate:

$$\sum_i \operatorname{vol}(\tilde{C}_i) = \sum_i (\kappa\sqrt{m})^m \operatorname{vol}(C_i) < (\kappa\sqrt{m})^m \epsilon_1 < \epsilon.$$

For the remainder of the proof, we shall assume that $\det(\Psi') \neq 0$ and that Ψ is injective on $\mathcal{O}$. By the Inverse Function Theorem (Theorem 1.8), Ψ is a homeomorphism from $\mathcal{O}$ to $\Psi(\mathcal{O})$.

Thus $\mathrm{bd}(\Psi(S)) = \Psi(\mathrm{bd}(S))$. We apply Lemma 1.18. Since S is Jordan measurable, $\mathrm{bd}(S)$ has content zero. By Assertion 2, $\Psi(\mathrm{bd}(S)) = \mathrm{bd}(\Psi(S))$ has content zero. This shows $\Psi(S)$ is Jordan measurable. This proves Assertion 3-a.

Assume that Ψ is linear. Suppose first S is a rectangle in C. As we can express Ψ as a composition of elementary row operations, it suffices to prove Assertion 3-b for S if Ψ is an elementary row operation. We use Fubini's Theorem to compute the volume as an iterated integral and reduce the computation to 2-dimensional questions. There are three cases:

Case 1: $\Psi(x, y) = (cx, y)$. It follows from Equation (1.5.a) that $\mathrm{vol}(\Psi(S)) = |c| \cdot \mathrm{vol}(S)$ so Assertion 3-b holds in this special case.

Case 2: $\Psi(x, y) = (y, x)$. It follows from Equation (1.5.a) that $\mathrm{vol}(\Psi(S)) = \mathrm{vol}(S)$ so Assertion 3-b holds in this special case.

Case 3: $\Psi(x, y) = (x + \lambda y, y)$. Let $S = [a, b] \times [c, d]$. Then

$$\Psi(S) := \{(x, y) : c \leq y \leq d, a + \lambda y \leq x \leq b + \lambda y\}.$$

We complete the proof of Assertion 3-b for rectangles by applying Fubini's Theorem:

$$\mathrm{vol}(\Psi(S)) = \int_c^d \int_{a+\lambda y}^{b+\lambda y} dx\, dy = \int_c^d \{(b + \lambda y) - (a + \lambda y)\} dy = (d - c)(b - a) = \mathrm{vol}(S).$$

Let S be an arbitrary Jordan measurable subset of C. Let $\mathcal{P}_n$ be the partition of C into n^m cubes of side c/n. Let $\mathcal{P}_n^i(S)$ be as defined in Equation (1.7.a). Then

$$\bigcup_{P \in \mathcal{P}_n^1(S)} P \subset S \subset \bigcup_{P \in \mathcal{P}_n^1(S) \cup \mathcal{P}_n^2(S)} P.$$

Let $\epsilon_1 > 0$ be given. Choose N so $n \geq N$ implies $\sum_{P \in \mathcal{P}_n^2(S)} \mathrm{vol}(P) < \epsilon_1$. We have:

$$
\begin{aligned}
|\det(\Psi')|\{\mathrm{vol}(S) - \epsilon_1\} &\leq |\det(\Psi')| \sum_{P \in \mathcal{P}_n : P \subset S} \mathrm{vol}(P) = \sum_{P \in \mathcal{P}_n : P \subset S} \mathrm{vol}(\Psi(P)) \\
&\leq \mathrm{vol}(\Psi(S)) \leq \sum_{P \in \mathcal{P}_n : P \cap S \neq \emptyset} \mathrm{vol}(\Psi(P)) \\
&= |\det(\Psi')| \sum_{P \in \mathcal{P}_n : P \cap S \neq \emptyset} \mathrm{vol}(P) \leq |\det(\Psi')|\{\mathrm{vol}(S) + \epsilon_1\}.
\end{aligned}
$$

We take the limit as $\epsilon_1 \to 0$ to establish Assertion 3-b for general Jordan measurable subsets of C.

Suppose Ψ satisfies the hypothesis of Assertion 3-c. Let $\Phi := T_0^{-1}\Psi - \mathrm{Id}$. Then:

$$\|\Phi'\| = \|T_0^{-1}\Psi' - \mathrm{Id}\| = \|T_0^{-1}\{\Psi' - T_0\}\| \leq \|T_0^{-1}\| \cdot \|\Psi' - T_0\| \leq \|T_0^{-1}\|\varepsilon.$$

Consequently, we may use Lemma 1.7 to estimate:

$$\|\Phi(x) - \Phi(y)\| \leq \epsilon \|T_0^{-1}\| \cdot \|x - y\| \quad \text{for all} \quad x, y \in C. \tag{1.7.f}$$

As above, let $\mathcal{P}_n$ be the partition of C into n^m cubes whose sides have length $s := c/n$. Let $P \in \mathcal{P}_n$. Let $x = (x^1, \ldots, x^m)$ and $y = (y^1, \ldots, y^m)$ be points of P. By Equation (1.7.f),

$$
\begin{aligned}
|\{(T_0^{-1}\Psi(x))^i - x^i\} - \{(T_0^{-1}\Psi(y))^i - y^i\}| &= |\Phi(x)^i - \Phi(y)^i| \\
&\leq \|\Phi(x) - \Phi(y)\| \\
&\leq \epsilon \|T_0^{-1}\| \cdot \|x - y\| \leq \|T_0^{-1}\|\sqrt{m}\epsilon s.
\end{aligned}
$$

Thus by the triangle inequality, we may estimate:

$$|(T_0^{-1}\Psi(x))^i - (T_0^{-1}\Psi(y))^i| \leq \|T_0^{-1}\|\sqrt{m}\epsilon s + |x^i - y^i| \leq (1 + \sqrt{m}\epsilon\|T_0^{-1}\|)s.$$

Consequently, $T_0^{-1}\Psi(P)$ is contained in a cube of side $(1 + \sqrt{m}\epsilon\|T_0^{-1}\|)s$ so:

$$\text{vol}(\Psi(P)) = |\det(T_0)| \cdot \text{vol}(T_0^{-1}\Psi(P)) \leq |\det(T_0)| \cdot (1 + \sqrt{m}\epsilon\|T_0^{-1}\|)^m \, \text{vol}(P).$$

This establishes Assertion 3-c if S is an element of the partition $\mathcal{P}_n$. More generally, let S be a Jordan measurable subset of C. Given $\epsilon_1 > 0$, we use Lemma 1.25 to find n so that

$$\text{vol}(S) \leq \sum_{P \in \mathcal{P}_n : P \cap S \neq \emptyset} \text{vol}(P) \leq \text{vol}(S) + \epsilon_1 \quad \text{for} \quad n \geq N.$$

We note that if A and B are Jordan measurable sets, then

$$\text{vol}(A) + \text{vol}(B) = \text{vol}(A \cup B) + \text{vol}(A \cap B).$$

In particular, if $A \cap B$ has content zero, then the volume is additive and we may ignore the possible overlap of $\Psi(P_1) \cap \Psi(P_2)$ for rectangles $P_i \in \mathcal{P}_n$ in what follows. We can use the inclusion $\Psi(S) \subset \bigcup_{P \in \mathcal{P}_n : P \cap S \neq \emptyset} \Psi(P)$ to see that:

$$
\begin{aligned}
\text{vol}(\Psi(S)) &\leq \text{vol}\left\{\cup_{P \in \mathcal{P}_n : P \cap S \neq \emptyset}\Psi(P)\right\} = \sum_{P \in \mathcal{P}_n : P \cap S \neq \emptyset} \text{vol}(\Psi(P)) \\
&\leq |\det(T_0)|(1 + \sqrt{m}\epsilon\|T_0^{-1}\|)^m \sum_{P \in \mathcal{P}_n : P \cap S \neq \emptyset} \text{vol}(P) \\
&\leq |\det(T_0)|\{1 + \sqrt{m}\epsilon\|T_0^{-1}\|\}^m\{\text{vol}(S) + \epsilon_1\}.
\end{aligned}
$$

The desired inequality now follows as ϵ_1 was arbitrary. $\qquad\square$

We can now establish a result which is dual in a certain sense to Theorem 1.9.

Theorem 1.27   *Let $\mathcal{O}$ be an open non-empty subset of $\mathbb{R}^m$ and let $H : \mathcal{O} \to \mathbb{R}^n$ be continuously differentiable. If $H(\mathcal{O})$ contains a non-empty open subset of $\mathbb{R}^n$, then $n \leq m$.*

Proof. Suppose to the contrary that $n > m$. Let $\Psi(x, y) := H(x)$ for $y \in \mathbb{R}^{n-m}$. Let S be a rectangle in $\mathbb{R}^m$. Then $S \times \{0\}$ has content zero in $\mathbb{R}^m$ and hence by Lemma 1.26, $\Psi(S)$ has content zero in $\mathbb{R}^n$. Consequently $H(S)$ has content zero in $\mathbb{R}^n$ so $H(\mathcal{O})$ has measure zero in $\mathbb{R}^n$ and cannot contain any non-empty open subset of $\mathbb{R}^n$. $\square$

1.7.1 CHANGE OF VARIABLE THEOREM. The following result will be central to our discussion of the generalized Stokes' Theorem:

Theorem 1.28   (Change of Variable Theorem).   *Let $\mathcal{O}$ be an open subset of $\mathbb{R}^m$. Let $\Phi : \mathcal{O} \to \mathbb{R}^m$ be continuously differentiable and injective with $\det(\Phi'(x)) \neq 0$ on $\mathcal{O}$. Let f be integrable on $\Phi(\mathcal{O})$ and let $\Phi^* f := f \circ \Phi$. Then $(\Phi^* f) \cdot |\det(\Phi')|$ is integrable on $\mathcal{O}$ and*

$$\int_{\Phi(\mathcal{O})} f = \int_{\mathcal{O}} (\Phi^* f) \cdot |\det(\Phi')|.$$

Proof. We shall divide the proof into several steps. We suppose $f \geq 0$ for the majority of the proof. In Step 1, we work locally and establish an inequality related to integrating over cubes. In Step 2, we extend this inequality to Jordan measurable sets and we reverse the roles of the domain and range to establish the Change of Variable Theorem for integrating over a cube; it is necessary to consider Jordan measurable sets rather than just cubes since $\Phi(\text{cube})$ need not, of course, be a cube. In Step 3, we pass to extended integrals using a partition of unity to complete the proof.

Step 1. Assume $f \geq 0$. Let C be a cube which is contained in $\mathcal{O}$. Let $\mathfrak{o}(f, x)$ be the oscillation of f that was defined in Equation (1.5.d). Set

$$D_\ell = \{x \in \Phi(C) : \mathfrak{o}(f, x) \geq \tfrac{1}{\ell}\} \quad \text{and} \quad D = \cup_{\ell \in \mathbb{N}} D_\ell = \{x \in \Phi(C) : \mathfrak{o}(f, x) > 0\}.$$

Since f is integrable on the Jordan measurable set $\Phi(C)$, D has measure zero and the sets D_ℓ have content zero. Applying Lemma 1.26 to Φ^{-1} yields $\Phi^{-1}(D_\ell)$ has content zero and hence $\Phi^{-1}(D) = \cup_{\ell \in \mathbb{N}} \Phi^{-1}(D_\ell)$ has measure zero. This implies $\Phi^* f$ is continuous almost everywhere on C. Since $\Phi(C)$ is compact, f is bounded on $\Phi(C)$ and thus $\Phi^* f$ is bounded on C. This shows that $\Phi^* f$ is integrable. We wish to show:

$$\int_{\Phi(C)} f \leq \int_C (\Phi^* f) \cdot |\det(\Phi')| \quad \text{if} \quad f \geq 0. \tag{1.7.g}$$

Let $\varepsilon > 0$ be given. Let $\kappa := \sup_{x \in C}\{\|\Phi'(x)^{-1}\|, \|\Phi'(x)\|\}$. Since Φ' is continuous on the cube C and since C is compact, Φ' is uniformly continuous on C. We may therefore choose $\epsilon > 0$ so that if $x_i \in C$ with $\|x_1 - x_2\| < \delta$, then $\|\Phi'(x_1) - \Phi'(x_2)\| < \varepsilon$. Choose a partition $\mathcal{P}$ of C which has mesh less than δ (i.e., if x_1 and x_2 are two points of any rectangle $P \in \mathcal{P}$, then $|x_1 - x_2| < \delta$). We then have

$$U\left((\Phi^* f) \cdot |\det \Phi'|, \mathcal{P}\right) \leq \varepsilon + \int_C (\Phi^* f) \cdot |\det \Phi'|.$$

Since $\|\Phi'(x_1) - \Phi'(x_2)\| < \epsilon$ for $x_1, x_2 \in P \in \mathcal{P}$, we apply Lemma 1.26 to see

$$\mathrm{vol}(\Phi(P)) \leq (1 + \sqrt{m}\kappa\epsilon)^m |\det(\Phi'(y))| \cdot \mathrm{vol}(P) \quad \text{for any} \quad y \in P\,.$$

This permits us to estimate

$$\mathrm{vol}(\Phi(P)) \leq (1 + \sqrt{m}\kappa\epsilon)^m \inf_{y \in P}(|\det(\Phi'(y))|) \cdot \mathrm{vol}(P)\,.$$

Clearly $f(z) \leq \sum_{P \in \mathcal{P}} M(f, \Phi(P))\chi_{\Phi(P)}(z)$ for any $z \in \Phi(C)$. Consequently:

$$\int_{\Phi(C)} f \leq \sum_{P \in \mathcal{P}} M(f, \Phi(P)) \int_{\Phi(C)} \chi_{\Phi(P)} = \sum_{P \in \mathcal{P}} \sup_{x \in P} \{f(\Phi(x))\} \cdot \mathrm{vol}(\Phi(P))$$

$$\leq (1 + \sqrt{m}\kappa\varepsilon)^m \sum_{P \in \mathcal{P}} \sup_{x \in P} \{(\Phi^* f)(x)\} \cdot \inf_{y \in P} \{|\det(\Phi'(y))|\} \cdot \mathrm{vol}(P)$$

$$= (1 + \sqrt{m}\kappa\varepsilon)^m \sum_{P \in \mathcal{P}} \sup_{x \in P} \left\{(\Phi^* f)(x) \cdot \inf_{y \in P} \{|\det(\Phi'(y))|\}\right\} \cdot \mathrm{vol}(P)$$

$$\leq (1 + \sqrt{m}\kappa\varepsilon)^m \sum_{P \in \mathcal{P}} \sup_{x \in P} \{(\Phi^* f)(x) \cdot |\det(\Phi'(x))|\} \cdot \mathrm{vol}(P)$$

$$= (1 + \sqrt{m}\kappa\varepsilon)^m U((\Phi^* f) \cdot |\det \Phi'|, \mathcal{P})$$

$$\leq (1 + \sqrt{m}\kappa\varepsilon)^m \left\{\varepsilon + \int_C (\Phi^* f) \cdot |\det \Phi'|\right\}\,.$$

Taking the limit as $\varepsilon \to 0$ establishes Equation (1.7.g).

Step 2. Let S be a Jordan measurable subset of C. Replacing f by $\chi_{\Phi(S)} f$ then yields:

$$\int_{\Phi(S)} f \leq \int_S (\Phi^* f) \cdot |\det(\Phi')|\,. \tag{1.7.h}$$

Assume additionally there exists a cube $\tilde{C}$ so $\Phi(S) \subset \tilde{C} \subset \Phi(\mathcal{O})$. Set

$$\tilde{\Phi} := \Phi^{-1}, \quad \tilde{f} := (\Phi^* f) \cdot |\det(\Phi')|, \quad \tilde{S} := \Phi(S)\,.$$

Applying Equation (1.7.h) to this setting yields:

$$\int_S (\Phi^* f) \cdot |\det(\Phi')| \leq \int_{\Phi(S)} \{(\Phi^{-1})^*[\Phi^* f]\} \cdot \{(\Phi^{-1})^*[|\det(\Phi')|]\} \cdot |\det(\Phi^{-1})'|\,. \tag{1.7.i}$$

As $\det(AB) = \det(A)\det(B)$, the chain rule implies:

$$\{(\Phi^{-1})^*[|\det(\Phi')|]\} \cdot |\det(\Phi^{-1})'| = |\det\{(\Phi' \circ \Phi^{-1}) \cdot (\Phi^{-1})'\}|$$
$$= |\det\{(\Phi \circ \Phi^{-1})'\}| = |\det(\mathrm{Id})| = 1\,.$$

Since $(\Phi^{-1})^*[\Phi^* f] = f \circ \Phi \circ \Phi^{-1} = f$, we may rewrite Equation (1.7.i) as:

$$\int_S (\Phi^* f) \cdot |\det(\Phi')| \le \int_{\Phi(S)} f \, .$$

Combining this inequality with Equation (1.7.h) shows

$$\int_{\Phi(S)} f = \int_S (\Phi^* f) \cdot |\det(\Phi')| \, . \tag{1.7.j}$$

Step 3. Assume that $f \ge 0$. Choose a cover $\{\mathcal{O}_\alpha\}$ of $\mathcal{O}$ by Jordan measurable open sets so that each $\mathcal{O}_\alpha$ is contained in some cube $C_\alpha \subset \mathcal{O}$ and so that each $\Phi(\mathcal{O}_\alpha)$ is contained in some cube $\tilde{C}_\alpha \subset \Phi(\mathcal{O})$. Let $\{\tilde{\phi}_n\}$ be a locally finite partition of unity on $\Phi(\mathcal{O})$ subordinate to the cover $\Phi(\mathcal{O}_\alpha)$. Then $\phi_n := \Phi^* \tilde{\phi}_n$ is a locally finite partition of unity on $\mathcal{O}$ subordinate to the cover $\mathcal{O}_\alpha$. Thus we may use Equation (1.7.j) to see

$$\int_{\Phi(\mathcal{O})} f = \sum_{n=1}^{\infty} \int_{\Phi(\mathcal{O}_{\alpha(n)})} \tilde{\phi}_n \cdot f = \sum_{n=1}^{\infty} \int_{\mathcal{O}_{\alpha(n)}} \phi_n \cdot (\Phi^* f) \cdot |\det(\Phi')|$$
$$= \int_{\mathcal{O}} (\Phi^* f) \cdot |\det(\Phi')| \, .$$

This establishes the Change of Variable Theorem if $f \ge 0$. Decomposing $f = f^+ - f^-$ as the difference of two positive functions then establishes the Change of Variable Theorem in complete generality. $\qquad\square$

CHAPTER 2

Manifolds

In Chapter 2, we will discuss the basic theory of smooth manifolds. In Section 2.1, we define what it means for M to be a smooth manifold or for M to be a submanifold of another manifold. We discuss immersions, fiber bundles, and vector bundles. In Section 2.2 we treat the tangent bundle, submersions, and the cotangent bundle. Section 2.3 deals with the exterior algebra and Stokes' Theorem. In Section 2.4, we present some applications of Stokes' Theorem.

2.1   SMOOTH MANIFOLDS

In this section, we present the basic foundational material concerning smooth manifolds that we shall need.

2.1.1 LOCALLY EUCLIDEAN SPACES. Let M be a locally compact metric space; the particular metric is not essential as only the underlying topology is relevant at this point. We say that M is *locally Euclidean* of dimension m if there exists an open cover $\{\mathcal{O}_\alpha\}$ of M and homeomorphisms ϕ_α from each $\mathcal{O}_\alpha$ to open subsets $\mathcal{U}_\alpha$ of $\mathbb{R}^m$. If $(x^1, \ldots, x^m)$ are the usual coordinates on $\mathbb{R}^m$, then we set

$$x^i_\alpha := x^i \circ \phi_\alpha \quad \text{to express} \quad \phi_\alpha = (x^1_\alpha, \ldots, x^m_\alpha) \quad \text{on} \quad \mathcal{O}_\alpha.$$

We call the pair $(\mathcal{O}_\alpha, \phi_\alpha)$ a *coordinate chart*. We have *transition functions* $\phi_{\alpha\beta} := \phi_\alpha \circ \phi_\beta^{-1}$;

$$\text{domain}\{\phi_{\alpha\beta}\} = \phi_\beta\{\mathcal{O}_\alpha \cap \mathcal{O}_\beta\} \subset \mathbb{R}^m \quad \text{and} \quad \text{range}\{\phi_{\alpha\beta}\} = \phi_\alpha\{\mathcal{O}_\alpha \cap \mathcal{O}_\beta\} \subset \mathbb{R}^m.$$

If all the functions $\{\phi_{\alpha\beta}\}$ are *diffeomorphisms* (i.e., if the functions $\phi_{\alpha\beta}$ are smooth and satisfy $\det(\phi'_{\alpha\beta}) \neq 0$), then M is said to be a *manifold* or to have a *smooth structure*. The collection $\{(\mathcal{O}_\alpha, \phi_\alpha)\}$ is said to be a *coordinate atlas*; the atlas can always be chosen to be maximal. We shall only work in the smooth setting and shall not consider locally Euclidean spaces with C^k structures or piecewise-linear structures. Thus the word "manifold" shall always mean "locally Euclidean with a smooth structure."

2.1.2 SMOOTH MAPS. If f is a map from a manifold M to $\mathbb{R}^n$, then f is *smooth* if

$$f \circ \phi_\alpha^{-1} : \mathcal{U}_\alpha \to \mathbb{R}^n \text{ is smooth for all } \alpha.$$

Let $C^\infty(M)$ be the infinite-dimensional real vector space of all smooth functions on M. If $\{(\mathcal{O}_\alpha, \phi_\alpha)\}$ is a coordinate atlas for a manifold M and if $\{(\tilde{\mathcal{O}}_\beta, \tilde{\phi}_\beta)\}$ is a coordinate atlas for a manifold $\tilde{M}$, then the Cartesian product

$$\{(\mathcal{O}_\alpha \times \tilde{\mathcal{O}}_\beta, \phi_\alpha \times \tilde{\phi}_\beta)\}$$

is a coordinate atlas which gives $M \times \tilde{M}$ a smooth structure. A function f from M to $\tilde{M}$ is said to be *smooth* if $\tilde{\phi}_\beta \circ f \circ \phi_\alpha^{-1}$ is a smooth map from $\mathcal{U}_\alpha$ to $\tilde{\mathcal{U}}_\beta$ for all α, β whenever $\tilde{\phi}_\beta \circ f \circ \phi_\alpha^{-1}$ is defined. If f is bijective, then f will be said to be a *diffeomorphism* if f and f^{-1} are both smooth; M and $\tilde{M}$ will be said to be *diffeomorphic* if there exists a diffeomorphism between them. There can be inequivalent smooth structures on the same underlying locally Euclidean metric space; there are manifolds which are homeomorphic but not diffeomorphic.

2.1.3 PULLBACK. Let $F : M \to N$ be smooth. For a function $f \in C^\infty(N)$, the *pullback* is defined by $F^*(f) := f \circ F \in C^\infty(M)$; F^* is a unital ring homomorphism from $C^\infty(N)$ to $C^\infty(M)$, i.e.,

$$F^*(f + g) = F^*(f) + F^*(g), \quad F^*(\mathbb{1}_N) = \mathbb{1}_M, \quad F^*(f \cdot g) = F^*(f) \cdot F^*(g),$$

where $\mathbb{1}$ denotes the constant function identically equal to 1. If G is a smooth map from a manifold L to a manifold M, we note that $(F \circ G)^* = G^* \circ F^*$ and $\mathrm{Id}^* = \mathrm{Id}$.

2.1.4 SUBMANIFOLDS. If W is a k-dimensional subspace of $\mathbb{R}^m$ and if $P \in \mathbb{R}^m$, then the translated subspace $A := W + P$ is said to be an *affine subspace* of $\mathbb{R}^m$ of dimension k. Let $\tilde{M}$ be a closed subset of a manifold M. We say that $\tilde{M}$ is a *submanifold* of M of dimension k if there exists a coordinate atlas $\{(\mathcal{O}_\alpha, \phi_\alpha)\}$ for M so that either $\mathcal{O}_\alpha \cap \tilde{M}$ is empty or $\phi_\alpha(\mathcal{O}_\alpha \cap \tilde{M}) = \mathcal{U}_\alpha \cap A^k$ where A^k is an affine subset of dimension k in $\mathbb{R}^m$. Setting $\tilde{\mathcal{O}}_\alpha := \mathcal{O}_\alpha \cap \tilde{M}$, and $\tilde{\phi}_\alpha := \phi_\alpha|_{\tilde{\mathcal{O}}_\alpha}$ then gives $\tilde{M}$ the structure of a manifold so the inclusion of $\tilde{M}$ in M is a smooth map.

   If $\mathcal{O}$ is an open subset of $\mathbb{R}^m$ and if V is a k-dimensional affine subspace of $\mathbb{R}^m$, then $\mathcal{O} \cap V$ is a k-dimensional submanifold of $\mathbb{R}^m$. There are, however, less trivial examples. Let $1 \le k < m$. Let M be an m-dimensional manifold and let $F : M \to \mathbb{R}^{m-k}$ be smooth. We say that F is *regular* at a point $P \in M$ if there exists a coordinate chart $(\mathcal{O}, \phi)$ with $P \in \mathcal{O}$ so that if we set $F_\phi := F \circ \phi^{-1}$ and if we set $Q = \phi(P)$, then $F'_\phi(Q)$ is a surjective map from $\mathbb{R}^m$ to $\mathbb{R}^{m-k}$. Note that this condition is independent of the particular coordinate chart chosen. We use Inverse Function Theorem to establish:

Theorem 2.1   *Let $1 \le k < m$. Let M be an m-dimensional manifold and let $F : M \to \mathbb{R}^{m-k}$ be smooth. Fix $c \in \mathbb{R}^{m-k}$ and let $M_c := F^{-1}(c)$ be the level set. Suppose F is regular at every point $P \in M_c$. Then M_c is a smooth submanifold of M of dimension k.*

Proof. Choose a coordinate chart $(\mathcal{O}, \phi)$ so $P \in \mathcal{O}$. By replacing M by $\mathcal{U} = \phi(\mathcal{O})$ and by replacing F by $F \circ \phi^{-1}$, we may reduce to the special case $F : \mathcal{U} \to \mathbb{R}^{m-k}$. Let $A_{ij} := \partial_{x^i} F_j$ for $1 \le i, j \le m - k$. Since $F'(P)$ is surjective, by permuting the coordinates if necessary, we may suppose that $\det(A) \ne 0$. Let $u = (x^1, \dots, x^{m-k})$ be the dependent variables and let $v = (x^{m-k+1}, \ldots, x^m)$ be the independent variables. Let $G(u, v) := (F(u, v), v)$. Since $\det(G'(P)) = \det(A) \neq 0$, G is an admissible change of coordinates and we have a new coordinate chart $(\mathcal{O}, \tilde{\phi})$ where $\tilde{\phi} := G \circ \phi$. Since $F(G(u, v)) = u$ is just projection on the first $m - k$ coordinates, this gives rise to a new coordinate atlas where the level sets of $F \circ G$ are linear subspaces of $\mathbb{R}^{m-k}$. $\qquad \square$

The following example is illustrative. Let $(V, \langle \cdot, \cdot \rangle)$ be an m-dimensional inner product space; we impose no restriction on the signature. Let $S_r(V, \langle \cdot, \cdot \rangle) := \{v \in V : \langle v, v \rangle = r\}$ be the *pseudo-sphere* of radius $r \neq 0$. Let $\{x^1, \ldots, x^m\}$ be the coordinates on V defined by an orthonormal basis $\{e^1, \ldots, e^m\}$ for V. Then $\langle x, x \rangle = \epsilon_1 (x^1)^2 + \cdots + \epsilon_m (x^m)^2$ where $\epsilon_i = \pm 1$. The derivative of the defining function is $(2\epsilon_1 x^1, \ldots, 2\epsilon_m x^m)$. This is non-zero on $V - \{0\}$. Thus $S_r(V, \langle \cdot, \cdot \rangle)$ is a smooth submanifold.

2.1.5 REGULAR COVERING SPACES. Let G be a finite group which acts smoothly on a manifold M without fixed points, i.e., $g \cdot x \neq x$ for $g \in G$ and for $x \in M$, and $g \neq \mathrm{Id}$. Let d be a metric on M. We give the orbit space $\tilde{M} := M/G$ the structure of a metric space by defining

$$\tilde{d}(xG, yG) := \min_{g_i \in G} d(g_1 x, g_2 y).$$

Choose local coordinate charts $(\mathcal{O}_\alpha, \phi_\alpha)$ on M so that $g\mathcal{O}_\alpha \cap \mathcal{O}_\alpha = \emptyset$ for $g \neq \mathrm{Id}$. The coordinate charts then descend to give $\tilde{M}$ the structure of a smooth manifold so that the natural projection $\pi : M \to \tilde{M}$ is a smooth map. The triple $(M, \pi, \tilde{M})$ is called a *regular covering projection*. We refer to Spanier [37] for further details.

2.1.6 REAL PROJECTIVE SPACE. There are several natural manifolds which arise as quotients of orthogonal actions by finite groups on the sphere. The *antipodal* map on the sphere is given by $a(x) = -x$. Since $a^2 = \mathrm{Id}$, $G := \{\mathrm{Id}, a\}$ is a group of order 2 that acts smoothly and without fixed points on the sphere $S^{m-1} \subset \mathbb{R}^m$. The quotient manifold $\mathbb{RP}^{m-1} := S^{m-1}/\mathbb{Z}_2$ is called *real projective space*. The map $\sigma : \xi \to \xi \cdot \mathbb{R}$ defines a map from the sphere to the set of lines through the origin in $\mathbb{R}^m$. Since $\sigma(\xi) = \sigma(\eta)$ if and only if $\xi = \pm \eta$. The map σ identifies $\mathbb{RP}^{m-1}$ with the space of lines through the origin in $\mathbb{R}^m$. If $\xi \in S^{m-1}$, then orthogonal projection on ξ is defined by $\pi_\xi : x \to (x, \xi)\xi$ where $(\cdot, \cdot)$ is the usual Euclidean inner product. Thus we may also identify $\mathbb{RP}^{m-1}$ with the space of orthogonal projections of rank 1 in $\mathrm{Hom}(\mathbb{R}^m, \mathbb{R}^m)$.

2.1.7 LENS SPACES AND SPHERICAL SPACE FORMS. If $m = 2n$ is even, we may regard S^{2n-1} as the unit sphere in $\mathbb{C}^n$. Identify the cyclic group $\mathbb{Z}_k$ of order k with the kth roots of unity in $\mathbb{C}$. Let $\vec{v} := (v_1, \ldots, v_n)$ be a collection of integers coprime to k. Let $\sigma_{\vec{v}}(\lambda) \cdot (z^1, \ldots, z^n) := (\lambda^{v_1} z^1, \ldots, \lambda^{v_n} z^n)$ define a fixed point free action $\sigma_{\vec{v}}$ of $\mathbb{Z}_k$ on S^{2n-1}. Let $L(k; \vec{v}) := S^{2n-1}/\sigma_{\vec{v}}(\mathbb{Z}_k)$; this is called a *lens space*. More generally the *spherical space forms* arise by setting $M = S^{m-1}/G$ where G is a finite subgroup of the orthogonal group $O(m)$ so that $\det(g - \mathrm{Id}) \neq 0$ for $g \neq \mathrm{Id}$. We refer to Wolf [41] for further details.

2.1.8 LIE GROUPS. We say that a manifold M is a *Lie group* if M is also a group such that the group multiplication $m(g, h) = g \cdot h$ from $M \times M$ to M is smooth and so that the map $g \to g^{-1}$ from M to M is smooth. Cramer's rule (Lemma 1.3) can be used to show that the *general linear group* $\mathrm{GL}(V)$ of invertible linear transformations of V is a Lie group. Sophus Lie initiated the study of continuous transformation groups and was a pioneer in this area.

Sophus Lie (1842–1899)

Let $(V, \langle \cdot, \cdot \rangle)$ be an inner product space. The associated *orthogonal group* is defined to be:

$$\mathcal{O}(V, \langle \cdot, \cdot \rangle) := \{ A \in \mathrm{GL}(V) : \langle Av, Aw \rangle = \langle v, w \rangle \text{ for all } v, w \in V \} .$$

We may verify that $\mathcal{O}(V, \langle \cdot, \cdot \rangle)$ is a Lie group as follows. If A belongs to $\mathrm{Hom}(V, V)$, then the adjoint A^*, which belongs to $\mathrm{Hom}(V, V)$, is characterized by the identity $\langle Av, w \rangle = \langle v, A^*w \rangle$ for all $v, w \in V$. Clearly $A \in \mathcal{O} = \mathcal{O}(V, \langle \cdot, \cdot \rangle)$ if and only if $A^*A = \mathrm{Id}$. It now follows that $\mathcal{O}$ is closed under composition, and that $A^* = A^{-1}$. Let $f(A) := A^*A$ map $\mathrm{Hom}(V, V)$ to the linear space of self-adjoint matrices $S := \{ B \in \mathrm{Hom}(V, V) : B = B^* \}$. Then $\mathcal{O} = f^{-1}(\mathrm{Id})$. Let $A \in \mathcal{O}$ and let $B \in S$. Let $\gamma(t) := A(1 + tB)$. Then

$$f(\gamma(t)) = (1 + tB^*)A^*A(1 + tB) = 1 + t(B^* + B) + O(t^2) = 1 + 2tB + O(t^2)$$

and thus $\partial_t f(\gamma(t))|_{t=0} = 2B$. This shows $f'(A)$ is surjective. We apply Theorem 2.1 to see $\mathcal{O}$ is a submanifold of $\mathrm{GL}(V)$. The group operation and the inverse map are the restriction of smooth maps on $\mathrm{GL}(V)$ and hence smooth on $\mathcal{O}$. Thus $\mathcal{O}$ is a Lie group. We will discuss more examples subsequently in Book II.

2.1.9 FIBER BUNDLES. We say that $F \to E \xrightarrow{\pi} M$ is a *fiber bundle* with fiber F over M if E is a manifold and if π is a smooth surjective map from E to M with $F = \pi^{-1}(P)$ where P is the base point of M. We assume there is a coordinate atlas $\{ (\mathcal{O}_\alpha, \phi_\alpha) \}$ for M and diffeomorphisms ψ_α from $\mathcal{O}_\alpha \times F$ to $\pi^{-1}(\mathcal{O}_\alpha)$ so that $P = \pi(\psi_\alpha(P, v))$ for any point $P \in \mathcal{O}_\alpha$ and any element v of the fiber F. Such a map is said to be *fiber preserving* since ψ_α defines a diffeomorphism from $\{ x \} \times F$ to the *fiber* $E_x := \pi^{-1}\{ x \}$. The *transition* or *gluing* functions are $\psi_{\alpha\beta} := \psi_\alpha^{-1} \circ \psi_\beta$. The maps $\psi_{\alpha\beta}$ are fiber preserving diffeomorphisms of $\{ \mathcal{O}_\alpha \cap \mathcal{O}_\beta \} \times F$. Note that we have the *cocycle condition*:

$$\psi_{\alpha\alpha} = \mathrm{Id} \text{ on } \mathcal{O}_\alpha \times F \quad \text{and} \quad \psi_{\alpha\beta} \circ \psi_{\beta\gamma} = \psi_{\alpha\gamma} \text{ on } (\mathcal{O}_\alpha \cap \mathcal{O}_\beta \cap \mathcal{O}_\gamma) \times F . \qquad (2.1.a)$$

2.1.10 IMMERSIONS. Let F be a smooth map from a manifold M to a manifold N. Let $(\mathcal{O}, \phi)$ be local coordinates near $P \in M$, and let $(\tilde{\mathcal{O}}, \tilde{\phi})$ be local coordinates near $F(P)$ in N. We say that F is an *immersion* if $(\tilde{\phi} \circ F \circ \phi^{-1})'(\phi(P))$ is an injective linear transformation for every $P \in M$; this condition is independent of the particular coordinate systems chosen. We say that F is an *embedding* if F is an immersion which is injective. Clearly if M is a submanifold of N, then the inclusion map is an embedding of M into N. The embedding is said to be a *proper embedding* if the inverse image of a compact set in N is a compact set in M.

2.1.11 THE KLEIN BOTTLE. This surface was first described by the German mathematician Felix Klein.

Christian Felix Klein (1849–1925)

Let $M = S^1$ be the unit circle. We regard $S^1 := [0, 2\pi]$ where we identify $0 \sim 2\pi$. We take $[0, 2\pi] \times S^1$ and glue $0 \times \lambda \sim 2\pi \times \bar{\lambda}$ (where $\bar{\lambda}$ is the complex conjugate) to define the *Klein bottle* $\mathbb{K}$. The natural projection $\pi : \mathbb{K} \to S^1$ is a fiber bundle with fiber S^1. The Klein bottle can be immersed, but not embedded in $\mathbb{R}^3$. We present below an immersion due to S. Dickson [14]; the surface is created in two pieces S_1 and S_2 which glue together smoothly but which intersect each other. Let $0 \le u \le \pi$ and $0 \le v \le 2\pi$. Set

$$S_1 \begin{pmatrix} x \\ y \\ z \end{pmatrix} = \begin{pmatrix} 6\cos(u)(1 + \sin(u)) + 4(1 - .5\cos(u))\cos(u)\cos(v) \\ 16\sin(u) + 4(1 - .5\cos(u))\sin(u)\cos(v) \\ 4(1 - .5\cos(u))\sin(v) \end{pmatrix},$$

$$S_2 \begin{pmatrix} x \\ y \\ z \end{pmatrix} = \begin{pmatrix} 6\cos(u + \pi)(1 + \sin(u + \pi)) + 4(1 - .5\cos(u + \pi))\cos(v + \pi) \\ 16\sin(u + \pi) \\ 4(1 - .5\cos(u + \pi))\sin(v) \end{pmatrix}.$$

We could also glue $0 \times \lambda \sim 2\pi \times \lambda$ to define the *torus* $\mathbb{T}^2 := S^1 \times S^1$. The natural projection $\pi : \mathbb{T}^2 \to S^1$ is also a fiber bundle with fiber S^1. We shall see presently that the torus $\mathbb{T}^2$ is orientable and the Klein bottle $\mathbb{K}$ is not orientable. Thus $\mathbb{T}^2$ is not diffeomorphic to $\mathbb{K}$. The torus can be embedded in $\mathbb{R}^3$. For $0 \le u \le 2\pi$ and $0 \le v \le 2\pi$, set:

$$\begin{pmatrix} x \\ y \\ z \end{pmatrix} = \begin{pmatrix} (3 + \cos(u))\cos(v) \\ (3 + \cos(u))\sin(v) \\ \sin(v) \end{pmatrix}, \quad 0 \le u, v \le 2\pi$$

Theorem 2.2   *Let M and $\tilde{M}$ be connected smooth manifolds.*

1. *If $F : M \to \tilde{M}$ is a proper embedding, then $F(M)$ is a submanifold of $\tilde{M}$ and F is a diffeomorphism from M to $F(M)$.*

2. **(Whitney Embedding Theorem [40])** *There exists a proper embedding of M into $\mathbb{R}^k$ for some k and, consequently, M is diffeomorphic to a closed proper submanifold of some Euclidean space.*

Proof. It suffices to establish Assertion 1 in the following special case; the assumption that F is proper then lets one pass from the local to the global setting without topological difficulties. Let $\mathcal{O}$ be an open subset of $\mathbb{R}^k$ and let $F : \mathcal{O} \to \mathbb{R}^m$ be smooth. Let $P \in \mathcal{O}$ and assume $F'(P)$ is an injective linear map from $\mathbb{R}^k$ to $\mathbb{R}^m$. If $k = m$, then the Inverse Function Theorem shows that F is a local diffeomorphism so we assume $k < m$. By making a linear change of coordinates on $\mathbb{R}^m$, we may assume that range$\{F'(P)\} = \mathbb{R}^k$ viewed as a subset of $\mathbb{R}^m$. Let $G(x, y) := (F(x), y)$ for $x \in \mathbb{R}^k$ and for $y \in \mathbb{R}^{m-k}$ define the germ of a map from $\mathbb{R}^m$ to itself. Then $G'(P)$ is injective and hence bijective as a map from $\mathbb{R}^m$ to itself. This provides the needed change of coordinates on $\mathbb{R}^m$ and establishes Assertion 1.

We shall assume that M is compact in proving Assertion 2; it is then automatic that the embedding is proper. Extending our proof to the general case requires dimension theory which is beyond the scope of this book. We refer to Munkres [32] for further details. Let $B_r(0)$ be the open ball of radius r about the origin. We can choose a coordinate atlas $\{(\mathcal{O}_\alpha, \phi_\alpha)\}$ for M so that ϕ_α is a diffeomorphism from $\mathcal{O}_\alpha$ to $B_3(0)$ and so that the open sets $\phi_\alpha^{-1}(B_1)$ still cover M. Since M is compact, there is a finite subcollection $\{(\mathcal{O}_i, \phi_i)\}$ for $1 \le i \le \ell$ so that $\{\phi_i^{-1}(B_1)\}$ covers M. Use Lemma 1.21 to find a smooth mesa function ψ so that $\psi = 1$ on $B_1(0)$ and so that ψ has compact support inside $B_2(0)$. Set:

$$\psi_i(x) := \psi(\phi_i(x)), \quad \phi_i^j(x) := \psi(\phi_i(x))x_i^j(x) \quad \text{if } x \in \mathcal{O}_i,$$
$$\psi_i(x) := 0, \qquad \phi_i^j(x) := 0 \qquad\qquad \text{if } x \notin \mathcal{O}_i.$$

Since the supports are contained in $\phi_i^{-1}(B_2(0))$, these functions are smooth on all of M. We define a smooth map Φ from M to $\mathbb{R}^{(m+1)\ell}$ by setting:

$$\Phi(x) := (\psi_1, \phi_1^1, \ldots, \phi_1^m, \ldots, \psi_\ell, \phi_\ell^1, \ldots, \phi_\ell^m).$$

Suppose $\Phi(P) = \Phi(Q)$. Choose ν so $P \in \phi_\nu^{-1}\{B_1(0)\}$. This implies $\psi_\nu(P) = 1$ and since $\Phi(P) = \Phi(Q)$, we also have $\psi_\nu(Q) = 1$. Consequently, P and Q both belong to $\mathcal{O}_\nu$. Since $\psi_\nu(P) = \psi_\nu(Q) = 1$, $x_\nu^j(P) = x_\nu^j(Q)$ for $1 \leq j \leq m$ so $P = Q$. This shows that Φ is injective. Furthermore, the coordinates $(x_\nu^1, \ldots, x_\nu^m)$ are among the components of Φ. Thus the Jacobian is injective so Φ is an embedding. Assertion 2 now follows from Assertion 1. $\qquad\square$

2.1.12 VECTOR BUNDLES. We now discuss vector bundles. Atiyah [4] or Karoubi [23] are good additional references. Let $F \to V \xrightarrow{\pi} M$ be a fiber bundle whose fiber F is a real or complex vector space of dimension r. If the transition functions $\psi_{\alpha\beta}$ preserve the vector space structure, then the fiber bundle is said to be a *vector bundle*. In this setting, we may regard $\psi_{\alpha\beta}$ as a smooth map from $\mathcal{O}_\alpha \cap \mathcal{O}_\beta$ to the general linear group $GL(F)$. Conversely, given a collection of smooth maps $\psi_{\alpha\beta}$ from $\mathcal{O}_\alpha \cap \mathcal{O}_\beta$ to $GL(F)$ which satisfy the cocycle condition given in Equation (2.1.a), we can recover the vector bundle in question by an appropriate identification of the disjoint union of the $\mathcal{O}_\alpha \times F$. The vector bundle is said to be a *line bundle* if $r = 1$. The *trivial bundle* of fiber dimension r is the Cartesian product $\mathbb{1}^r = M \times \mathbb{R}^r$.

2.1.13 TRIVIALIZATIONS AND BUNDLE MAPS. Let us consider $F \to V \xrightarrow{\pi} M$ and $\tilde{F} \to \tilde{V} \xrightarrow{\tilde{\pi}} M$ two vector bundles over M modeled on vector spaces F and $\tilde{F}$. By an abuse of notation we shall in the future simply say V and $\tilde{V}$ are vector bundles over M where the projections π and $\tilde{\pi}$ are implicit. A *trivialization* of V over an open subset $\mathcal{O}$ of M is a fiber preserving diffeomorphism $\psi : \mathcal{O} \times F \to \pi^{-1}(\mathcal{O})$ which is linear on the fibers; V is said to be *trivial* over $\mathcal{O}$ if there exists a trivialization. The ψ_α give trivializations of V over the open sets $\mathcal{O}_\alpha$ of the cover. We say that Ξ is a *bundle map* from V to $\tilde{V}$ if Ξ is a smooth map from V to $\tilde{V}$ which is a linear map from the fibers V_P to the fibers $\tilde{V}_P$ for each $P \in M$. We say Ξ is a *bundle isomorphism* if Ξ is a diffeomorphism. We refer to Atiyah [4] for the proof of the following:

Lemma 2.3 If $(\mathcal{O}, \phi)$ is a coordinate chart so that $\phi(\mathcal{O})$ is an open convex subset of $\mathbb{R}^m$, then any vector bundle over $\mathcal{O}$ is trivial. If Ξ is a bundle map from V to $\tilde{V}$ such that $\text{Rank}(\Xi_x)$ is constant on M, then $\ker\{\Xi\}$ is a subbundle of V and $\text{range}\{\Xi\}$ is a subbundle of $\tilde{V}$.

2.1.14 SECTIONS, FRAMES, TRANSITION FUNCTIONS, AND FIBER METRICS. Let V be a vector bundle over M of fiber dimension r. A *section* to V over an open subset $\mathcal{O}$ of M is a smooth map s from $\mathcal{O}$ to V so that $\pi(s(P)) = P$ for all $P \in \mathcal{O}$. Fix a basis $\{e_1, \ldots, e_r\}$ for the model vector space F. Let $\{(\mathcal{O}_\alpha, \phi_\alpha)\}$ be a coordinate atlas for M so that V is trivial over each $\mathcal{O}_\alpha$. Over each $\mathcal{O}_\alpha$, we have local sections $s_{i,\alpha}(x) := \psi_\alpha(x, e_i)$. The collection $\vec{s}_\alpha := (s_{1,\alpha}, \ldots, s_{r,\alpha})$ gives a *local frame* for V over $\mathcal{O}_\alpha$, i.e., $\{s_{1,\alpha}, \ldots, s_{r,\alpha}\}$ is a basis for the *fiber* V_x of V over x. Conversely, given such a frame over an open set $\mathcal{O}$, we obtain a *local trivialization* of V over $\mathcal{O}$ by setting $\phi_\mathcal{O}(x, \lambda) := \lambda_1 s_1(x) + \cdots + \lambda_r s_r(x)$. By an abuse of notation, let $C^\infty(V)$ be the set of smooth sections to V. Fiberwise addition and multiplication give $C^\infty(V)$ the structure of module over the ring of smooth functions on M.

Let V and W be vector bundles over M. We choose a coordinate atlas $\{(\mathcal{O}_\alpha, \phi_\alpha)\}$ so that V and W are trivial over each $\mathcal{O}_\alpha$ and are defined by transition functions $\Psi^V_{\alpha\beta}$ and $\Psi^W_{\alpha\beta}$, respectively. The direct sum vector bundle $V \oplus W$ and the tensor product vector bundle $V \otimes W$ are defined, respectively, by $\Psi^V_{\alpha\beta} \oplus \Psi^W_{\alpha\beta}$ and $\Psi^V_{\alpha\beta} \otimes \Psi^W_{\alpha\beta}$. The dual vector bundle V^* is defined using the induced dual transition functions on the dual vector space $\{\Psi^V_{\alpha\beta}{}^*\}^{-1}$ and the vector bundle $\mathrm{Hom}(V, W) := V \otimes W^*$ is defined by $\Psi^V_{\alpha\beta} \otimes \Psi^W_{\alpha\beta}{}^*$. Other bundles may be defined similarly. If V has fiber dimension r and if W has fiber dimension s, then $V \oplus W$ has fiber dimension $r + s$, $V \otimes W$ has fiber dimension $r \cdot s$, and V^* has fiber dimension r.

A *fiber metric* on a vector bundle V is a positive definite inner product $(\cdot, \cdot)$ on the fibers V_x which varies smoothly. Given such a metric, the *unit sphere bundle* $S(V)$ and the *unit disk bundle* $D(V)$ are defined by:

$$S(V) = \{\xi \in V : (\xi, \xi) = 1\} \quad \text{and} \quad D(V) = \{\xi \in V : (\xi, \xi) \le 1\}.$$

Note that $S(V)$ is the boundary of $D(V)$. By using the *Gram–Schmidt process*, one can choose orthonormal frames; we then have $\psi_{\alpha\beta}$ is an orthogonal matrix in the real setting and a unitary matrix in the complex setting.

2.1.15 THE TORUS AND THE MÖBIUS STRIP. The torus $\mathbb{T}^2$ is the unit circle bundle of the trivial bundle $\mathbb{1}^2$ over S^1 and the Möbius strip $\mathbb{M} := [0, 2\pi] \times [-1, 1]/\sim$ is defined by gluing $0 \times \lambda \sim 2\pi \times (-\lambda)$. One embeds the Möbius strip $\mathbb{M}$ in $\mathbb{R}^3$ by setting:

$$x(u, v) = (3 + u \cos(.5v)) \cos(v), \quad y(u, v) = (3 + u \cos(.5v)) \sin(v), \quad z(u, v) = u \sin(.5v)$$

for $u \in [-1, 1]$ and $v \in [0, 2\pi]$. One glues $(u, 0)$ to $(1 - u, \pi)$ to obtain:

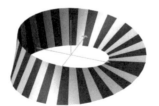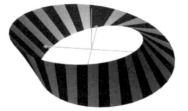

Replacing $[-1, 1]$ by $\mathbb{R}$ yields a non-trivial line bundle $\mathbb{L}$ over the circle. Define an open cover of the circle by setting: $\mathcal{O}_1 := S^1 - \{(1, 0)\}$ and $\mathcal{O}_2 := S^1 - \{(-1, 0)\}$. Then $\mathcal{O}_1 \cap \mathcal{O}_2$ consists of two disjoint open sets. The gluing function is $+1$ on one and -1 on the other of these sets. The Möbius strip is the unit disk bundle of $\mathbb{L}$ and the Klein bottle is the unit circle bundle of $\mathbb{1} \oplus \mathbb{L}$ over the circle. The transition function for $\mathbb{L} \otimes \mathbb{L}$ is given by $(\pm 1)^2 = 1$ and thus $\mathbb{L} \otimes \mathbb{L} = \mathbb{1}^2$ is the trivial line bundle. One can also show that $\mathbb{L} \oplus \mathbb{L}$ is the trivial 2-plane bundle although this takes more work. We have $\mathbb{L}^*$ has transition functions $\{(+1)^{-1}, (-1)^{-1}\}$; $\mathbb{L}$ is isomorphic to $\mathbb{L}^*$.

2.1.16 TENSORS. Let V and W be vector bundles over a manifold M. We say that a linear map T from $C^\infty(V)$ to $C^\infty(W)$ is a *tensor* or is *tensorial* if T is a $C^\infty(M)$ *module homomorphism*; that means that $T(fs) = fT(s)$ for $s \in C^\infty(V)$ and $f \in C^\infty(M)$.

Lemma 2.4 Let T be a tensorial map from $C^\infty(V)$ to $C^\infty(W)$. Then T arises from a bundle map from V to W.

Proof. Let $\{s_1, \ldots, s_r\}$ be a local frame field for V over a coordinate neighborhood $(\mathcal{O}, (x^1, \ldots, x^m))$. It suffices to show that if $s \in C^\infty(V)$ satisfies $s(P) = 0$, then we have that $(T(s))(P) = 0$. Express $s(x) = \sum_i f^i(x)s_i(x)$. Use Lemma 1.21 to find a smooth function ψ with compact support in $\mathcal{O}$ so ψ is identically 1 near P. Choose ξ with compact support in $\mathcal{O}$ so that ξ is identically 1 on the support of ψ. Then ξs_i and ψf^i extend smoothly to all of M. Furthermore $\xi\psi = \psi$. We compute:

$$
\begin{aligned}
(T(s))(P) &= (T((1-\psi)s))(P) + (T(\psi s))(P) = (1-\psi)(P)(T(s))(P) + (T(\xi\psi s))(P) \\
&= 0 + \sum_i (T(\psi f^i \xi s_i))(P) = \sum_i (\psi f^i(P)) \cdot (T(\xi s_i))(P) = 0.
\end{aligned}
$$

Suppose given $X(P)$ in the fiber of V over P. Expand $X(P) = \sum_i a^i s_i(P)$. Define a smooth global section to V by defining $\check{X} := \sum_i \psi a^i s_i(P)$. Then define a map $T(P)$ from V_P to W_P by setting $T(P) := \sum_i (T(\psi a^i s_i)))(P)$. This is independent of the choices made. We see that T is smooth by noting that $T(\psi a^i s_i) = \sum_i a^i s_i$ where $\psi = 1$. $\qquad\square$

2.2   THE TANGENT AND COTANGENT BUNDLES

In this section, we shall adopt the *Einstein convention* and sum over repeated Roman indices; indices α, β, and γ will index coordinate charts and are not summed over. Let $\{(\mathcal{O}_\alpha, \phi_\alpha)\}$ be a coordinate atlas for M. Express $\phi_\alpha = (x_\alpha^1, \ldots, x_\alpha^m)$ where the $\{x_\alpha^i := \phi_\alpha^* x^i\}$ are the coordinate functions on $\mathcal{O}_\alpha$. Let $f \in C^\infty(M)$. Then the *pullback* $(\phi_\alpha^{-1})^* f := f \circ \phi_\alpha^{-1}$ is a smooth function on $\mathcal{U}_\alpha = \phi_\alpha(\mathcal{O}_\alpha)$ and we define $\partial_{x_\alpha^i} f := \partial_{x^i}\{(\phi_\alpha^{-1})^* f\}$. By the Chain Rule,

$$
\partial_{x_\alpha^i} = \frac{\partial x_\beta^j}{\partial x_\alpha^i} \cdot \partial_{x_\beta^j} \quad \text{on} \quad \mathcal{O}_\alpha \cap \mathcal{O}_\beta. \tag{2.2.a}
$$

The transition functions $\Psi_{\alpha\beta}$ are given by the *Jacobian matrix* $\Psi_{\alpha\beta}{}^j{}_i := \partial_{x_\alpha^i} x_\beta^j$ and we have the *cocycle condition* of Equation (2.1.a):

$$
\Psi_{\alpha\beta}{}^j{}_i \cdot \Psi_{\beta\gamma}{}^k{}_j = \Psi_{\alpha\gamma}{}^k{}_i \quad \text{on} \quad \mathcal{O}_\alpha \cap \mathcal{O}_\beta \cap \mathcal{O}_\gamma.
$$

We denote the resulting vector bundle over M by the *tangent bundle* TM; the fiber $T_P M$ is called the *tangent space* at P. Given a local system of coordinates $\vec{x}_\alpha = (x_\alpha^1, \ldots, x_\alpha^m)$, the $\{\partial_{x_\alpha^i}\}$ form a local frame. A global section to TM is said to be a *tangent vector field*. By an abuse of notation, we let $C^\infty(TM)$ be the infinite-dimensional vector space consisting of all smooth tangent vector fields. This is a module under pointwise scalar multiplication over the space of smooth functions on M.

2.2.1 DERIVATIONS. We say that a linear map $\theta : C^\infty(M) \to \mathbb{R}$ is a *derivation based at P* if

$$\theta(fg) = \theta(f)g(P) + f(P)\theta(g) \quad \text{for all} \quad f, g \in C^\infty(M).$$

Similarly we say that a linear map $\Theta : C^\infty(M) \to C^\infty(M)$ is a *derivation* if

$$\Theta(fg) = \Theta(f)g + f\Theta(g) \quad \text{for all} \quad f, g \in C^\infty(M).$$

If $X = a^i \partial_{x^i} \in T_P M$ is a tangent vector at P, let $X(f)(P) := a^i \partial_{x^i} f(P)$ where we suppress the role of the coordinate chart index 'α' to simplify the notation. This is a derivation based at P. Similarly if X is a global tangent vector field, then we obtain a derivation of $C^\infty(M)$.

Lemma 2.5   If θ is a derivation of $C^\infty(M)$ which is based at $P \in M$, then there exists a unique tangent vector $X \in T_P M$ so that $\theta(f) = X(f)(P)$ for any $f \in C^\infty(M)$. If Θ is a derivation of $C^\infty(M)$, then there exists a unique tangent vector field X so $\theta(f) = X(f)$.

Proof. Let B_r be the open ball of radius r about the origin in $\mathbb{R}^m$. Let θ be a derivation of $C^\infty(M)$ based at $P \in M$. Choose a coordinate system $(\mathcal{O}, \phi)$ so $P \in \mathcal{O}$, so ϕ is a diffeomorphism from $\mathcal{O}$ to B_3, and so $\phi(P) = 0$. To avoid notational complexities, we identify $\mathcal{O}$ with B_3, we identify P with 0, and we ignore the role of ϕ. Let $\mathbb{1}$ be the constant function identically equal to 1. We observe

$$\theta(\mathbb{1}) = \theta(\mathbb{1} \cdot \mathbb{1}) = \theta(\mathbb{1}) \cdot \mathbb{1}(0) + \mathbb{1}(0) \cdot \theta(\mathbb{1}) = 2\theta(\mathbb{1}).$$

Consequently, $\theta(\mathbb{1}) = 0$. Fix $f \in C^\infty(M)$. Let $x \in B_3$. Set

$$g_i(x) := \int_0^1 (\partial_{x^i} f)(tx)dt.$$

Note that $g_i(0) = (\partial_{x^i} f)(0)$. We have

$$f(x) - f(0) = \int_0^1 \partial_t \{f(tx)\}dt = \int_0^1 x^i \{(\partial_{x^i} f)(tx)\}dt = x^i g_i(x). \tag{2.2.b}$$

Use Lemma 1.21 to find a smooth mesa function η taking values in $[0, 1]$, so that $\eta = 1$ on B_1 and so $\eta = 0$ on B_2^c. Note that η^2 also is a mesa function. Since $\eta^2 g_i$ and $\eta^2 x^i$ have support in B_2, we can extend them to be zero on B_2^c and regard these functions as belonging to $C^\infty(M)$. We decompose $f = f_0 + f_1 + f_2$ where the $f_i \in C^\infty(M)$ are defined by

$$f_0 := f(0)\mathbb{1}, \quad f_1 = (1 - \eta^2)(f - f_0), \quad f_2 = \eta^2(f - f_0).$$

We consider the tangent vector $X := \theta(\eta x^i) \cdot \partial_{x^i} \in T_P M$. Because $\Theta(\mathbb{1}) = 0$, we have
$$\theta(f_0) = \theta(f(0)\mathbb{1}) = 0.$$

Because $(1 - \eta^2)(0) = 0$ and $(f - f_0)(0) = 0$,

$$\theta(f_1) = \theta((1 - \eta^2)(f - f_0))$$
$$= \theta(1 - \eta^2) \cdot (f - f_0)(0) + (1 - \eta^2)(0) \cdot \theta(f - f_0) = 0 + 0 = 0.$$

Equation (2.2.b) implies that $\eta^2(f - f_0) = \eta^2(x^i g_i) = (\eta x^i)(\eta g_i)$. Consequently

$$\theta(f_2) = \theta(\eta^2(f - f_0)) = \theta((\eta x^i)(\eta g_i)) = \theta(\eta x^i) \cdot (\eta g_i)(0) + (\eta x^i)(0) \cdot \theta(\eta g_i)$$
$$= \theta(\eta x^i) g_i(0) = \theta(\eta x^i) \cdot (\partial_{x^i} f)(0) = (Xf)(0).$$

This shows $\theta(f) = (Xf)(0)$ and represents θ as a tangent vector; since the coefficients are $\theta(\eta x^i)$, the representation is unique. If Θ is a derivation of $C^\infty(M)$, then we define a derivation based at P by setting $\theta_P(f) := \Theta(f)(P)$. If $X = \theta(\eta^2 x^i)\partial_{x^i}$, then $\Theta(f) = X(f)$ in the coordinate chart $\mathcal{O}$; since X is globally defined, $\Theta = X$. □

2.2.2 THE LIE BRACKET. Let X and Y be vector fields on a manifold M. By Lemma 2.5, we can identify X and Y with derivations of $C^\infty(M)$. The commutator $[X, Y]$, characterized by the identity $[X, Y](f) = (XY - YX)f$ is a linear map from $C^\infty(M)$ to $C^\infty(M)$ which is called the *Lie bracket*. We verify that it is a derivation by computing:

$$[X, Y](fg) = X(Y(fg)) - Y(X(fg)) = X\{Y(f)g + fY(g)\} - Y\{X(f)g + fX(g)\}$$
$$= X(Y(f))g + Y(f)X(g) + X(f)Y(g) + fX(Y(g))$$
$$- Y(X(f))g - X(f)Y(g) - Y(f)X(g) - fY(X(g))$$
$$= [X, Y](f) \cdot g + f \cdot ([X, Y](g)).$$

If $X = a^i \partial_{x^i}$ and $Y = b^j \partial_{x^j}$, then $[X, Y] = \{a^j \partial_{x^j}(b^i) - b^j \partial_{x^j}(a^i)\}\partial_{x^i}$. The Lie bracket satisfies the *Jacobi identity*:

$$[[X, Y], Z] + [[Y, Z], X] + [[Z, X], Y] = 0.$$

2.2.3 THE FLOW OF A VECTOR FIELD. If θ is a derivation of $C^\infty(M)$ which is based at P and if $F : M \to N$ is smooth, define $(F_*\theta)(f) = \theta(F^*f)$ for $f \in C^\infty(M)$; this is a derivation based at $F(P)$. The *pushforward* is a linear map F_* from $T_P M$ to $T_{F(P)}N$ given locally by:

$$F_*(\partial_{x^i}) = \partial_{x^i} y^j \cdot \partial_{y^j}; \quad (G \circ F)_* = G_* \circ F_* \quad \text{and} \quad \text{Id}_* = \text{Id}. \tag{2.2.c}$$

We say that a map $\Phi : M \times \mathbb{R} \to M$ has the *semigroup property* if $\Phi(y, 0) = y$ for all y and if $\Phi(\Phi(y, t), s) = \Phi(y, t + s)$ for all y, t, and s.

Lemma 2.6 Let X be a smooth vector field on a compact manifold M. There exists a unique smooth map $\Phi : M \times \mathbb{R} \to M$, called the flow of X, which has the semigroup property so that $\Phi_*(\partial_t) = X(\Phi(y, t))$ for all y, t.

Proof. Suppose first that X is a smooth vector field on an open subset $\mathcal{O}$ of $\mathbb{R}^m$. Expand $X = a^i \partial_{x^i}$ to regard $X = (a^1(x), \dots, a^m(x))$ as a smooth map from $\mathcal{O}$ to $\mathbb{R}^m$; this is, of course, just the canonical identification of $T\mathcal{O} = \mathcal{O} \times \mathbb{R}^m$. Let C be a compact subset of $\mathcal{O}$. The Fundamental Theorem of Ordinary Differential Equations shows there exists $\epsilon > 0$ and a unique smooth map $\Psi : C \times [-\epsilon, \epsilon] \to \mathbb{R}^m$ so that

$$(\partial_t \Psi)(y, t) = X(\Psi(y, t)) \quad \text{on} \quad C \times [-\epsilon, \epsilon].$$

This map has the semigroup property for $|t| + |s| \le \epsilon$. Let $\{(\mathcal{O}_\alpha, \phi_\alpha)\}$ be a coordinate atlas on M so that the corresponding subsets $U_\alpha \subset \mathbb{R}^m$ are all the open ball $B_2(0)$ of radius 2 about the origin. Let $C_\alpha := \phi_\alpha^{-1}\{\bar{B}_1(0)\}$. Since $\{\text{int}(C_\alpha)\}$ is a cover of M, we can choose a finite sub-cover $\{\text{int}(C_1), \dots, \text{int}(C_\ell)\}$ of M. The argument quoted above permits us to construct flows Ψ_i for X with domains $C_\nu \times [-\epsilon_\nu, \epsilon_\nu]$ which take values in $\mathcal{O}_\nu$. One sets $\epsilon = \min(\epsilon_1, \dots, \epsilon_\nu)$. The uniqueness of the flow shows that $\Psi_i = \Psi_j$ on their common domain. This constructs a flow $\Psi : M \times [-\epsilon, \epsilon] \to M$ satisfying the semigroup property for $|s| + |t| < \epsilon$. We iterate the process and use the semigroup property to construct the desired flow. $\qquad\square$

2.2.4 THE FROBENIUS THEOREM. We begin our discussion of the Frobenius Theorem with the following technical result:

Lemma 2.7 Let $\{X_1, \dots, X_k\}$ be smooth vector fields on M satisfying $[X_i, X_j] = 0$ for all i, j and so that $\{X_1(P), \dots, X_k(P)\}$ are linearly independent at a point P of M.

1. We can choose local coordinates $\vec{x} = (x^1, \dots, x^m)$ so that $\partial_{x^1} = X_1$ near P.

2. We have $\Phi_t^{X_i} \Phi_s^{X_j} = \Phi_s^{X_j} \Phi_t^{X_i}$ near P.

3. We can find local coordinates on M centered at P so $X_i = \partial_{x^i}$ for $1 \le i \le k$.

Proof. Let $(x^1, \dots, x^m)$ be local coordinates on M centered at P. Make a linear change of coordinates, to assume $X_1(P) = \partial_{x^1}(P)$. Let $T(x^1, x^2, \dots, x^m) := \Phi_{x^1}^{X_1}(0, x^2, \dots, x^m)$. This is a smooth map and $T_*(0) = \text{Id}$. Thus T is an admissible change of coordinates and clearly in this new coordinate system, $\partial_{x^1} = X_1$. This proves Assertion 1.

By Assertion 1, we may assume the coordinates are chosen so $X_1 = \partial_{x^1}$. Sum over repeated indices to expand $X_2 = a^i(\vec{x}) \partial_{x^i}$. Since the bracket of X_1 and X_2 is zero, $a^i(\vec{x})$ is independent of x^1. Let $(e_1, \dots, e_m)$ be the standard basis for $\mathbb{R}^m$:

$$e_1 = (1, 0, \dots, 0), \qquad e_2 = (0, 1, 0, \dots, 0), \qquad \dots.$$

Then a point $\vec{x} \in \mathbb{R}^m$ can be represented as $\vec{x} = x^1 e_1 + \dots + x^m e_m$. Consequently the flow for X_1 is the linear map $\Psi_s^{X_1} : e_1 \to e_1 + s$ while $\Phi_s^{X_1} : e_i \to e_i$ for $i \ge 2$ or, equivalently, $\Psi_s^{X_1}(\vec{x}) =$

$\vec{x} + se_1$. Let

$$\Psi_{t,s}(\vec{x}) := \Psi_s^{X_1} \left\{ \Psi_t^{X_2}(\vec{x}) \right\} = \Psi_t^{X_2}(\vec{x}) + se_1,$$
$$\Theta_{t,s}(\vec{x}) := \Psi_t^{X_2} \left\{ \Psi_s^{X_1}(\vec{x}) \right\} = \Psi_t^{X_2}(\vec{x} + se_1).$$

Fix s. We have the initial conditions $\Psi_{0,s}(\vec{x}) = \vec{x} + se_1$ and $\Theta_{0,s}(\vec{x}) = \vec{x} + se_1$. Since the coefficients a^i are independent of x^1, we have the evolution equations:

$$\partial_t \Psi_{t,s}(\vec{x}) = \partial_t \Psi_t^{X_2}(\vec{x}) = X_2(\Psi_t^{X_2}(\vec{x})) = X_2(\Psi_t^{X_2}(\vec{x}) + se_1) = X_2(\Psi_{t,s}(\vec{x})),$$
$$\partial_t \Theta_{t,s}(\vec{x}) = X_2(\Psi_t^{X_2}(\vec{x} + se_1)) = X_2(\Theta_{t,s}(\vec{x})).$$

Assertion 2 now follows from the Fundamental Theorem of Ordinary Differential Equations for $i = 1$ and $j = 2$ while general case then follows by permuting the indices. We now make a linear change of coordinates to assume $X_i(P) = \partial_{x^i}$ for $1 \le i \le k$. We define

$$T(x^1, \ldots, x^k, x^{k+1}, \ldots, x^m) = (\Phi_{x^1}^{X_1} \circ \cdots \circ \Phi_{x^k}^{X_k})(0, \ldots, 0, x^{k+1}, \ldots, x^m).$$

$X_1 = \partial_{x^1}$ in these coordinates. As we can commute the $\Phi's$, $X_i = \partial_{x^i}$ for $1 \le i \le k$. □

Theorem 2.8 (Frobenius [16]). *Let V be a subbundle of TM of dimension k. The following conditions are equivalent and if either is satisfied, then V is said to be integrable.*

1. *If $X \in C^\infty(V)$ and $Y \in C^\infty(V)$, then $[X, Y] \in C^\infty(V)$.*

2. *There are local coordinates $(x^1, \ldots, x^m)$ around any point $P \in M$ so that the collection $\{\partial_{x^1}, \ldots, \partial_{x^k}\}$ forms a local frame for V.*

Proof. Suppose Assertion 2 holds. Let $X \in C^\infty(V)$ and $Y \in C^\infty(V)$. Fix $P \in M$. Choose local coordinates $(x^1, \ldots, x^m)$ so $V = \text{span}\{\partial_{x^1}, \ldots, \partial_{x^k}\}$. Expand

$$X = \sum_{i \le k} a^i \partial_{x^i} \quad \text{and} \quad Y = \sum_{j \le k} b^j \partial_{x^j}.$$

We may show that Assertion 2 implies Assertion 1 by computing:

$$[X, Y] = \sum_{i,j \le k} \left\{ a^i \partial_{x^i} b^j - b^i \partial_{x^i} a^j \right\} \cdot \partial_{x^j} \in C^\infty(V).$$

Suppose Assertion 1 holds. Let $\{X_1, \ldots, X_k\}$ be smooth sections to V which are linearly independent at P. They then form a local frame for V near P. Expand

$$X_i = a^{i1} \partial_{x^1} + \cdots + a^{im} \partial_{x^m}.$$

Since $X_1(P) \neq 0$, we have $a^{1i}(P) \neq 0$ for some i. By renumbering, we assume $a^{11}(P) \neq 0$. By replacing X_1 by $(a^{11})^{-1}X_1$ and shrinking the neighborhood under consideration, we may assume $a^{11} = 1$. By replacing X_i by $X_i - a^{i1}X_1$, we can normalize the frame so that

$$
\begin{aligned}
X_1 &= 1 \cdot \partial_{x^1} &+& \ a^{12}\partial_{x^2} &+& \ \dots &+& \ a^{1m}\partial_{x^m}, \\
X_2 &= 0 \cdot \partial_{x^1} &+& \ a^{22}\partial_{x^2} &+& \ \dots &+& \ a^{2m}\partial_{x^m}, \\
&\quad \dots && \dots && \dots && \dots
\end{aligned}
$$

By permuting the indices, we may assume $a^{22} \neq 0$. Multiplying X_2 by $(a^{22})^{-1}$ and shrinking the neighborhood permits us to assume $a^{22} = 1$. Replacing X_i by $X_i - a^{i2}X_2$ for $i \neq 2$ then permits us to assume $a^{i2} = 0$ for $i \neq 2$ and thus our frame has the form:

$$
\begin{aligned}
X_1 &= 1 \cdot \partial_{x^1} &+& \ 0 \cdot \partial_{x^2} &+& \ a^{13}\partial_{x^3} &+& \ \dots &+& \ a^{1m}\partial_{x^m}, \\
X_2 &= 0 \cdot \partial_{x^1} &+& \ 1 \cdot \partial_{x^2} &+& \ a^{23}\partial_{x^3} &+& \ \dots &+& \ a^{2m}\partial_{x^m}, \\
X_3 &= 0 \cdot \partial_{x^1} &+& \ 0 \cdot \partial_{x^2} &+& \ a^{33}\partial_{x^3} &+& \ \dots &+& \ a^{3m}\partial_{x^m}, \\
&\quad \dots && \dots && \dots && \dots
\end{aligned}
$$

Continue in this way to choose a local frame so $X_i = \partial_{x^i} + \sum_{\ell > k} a^{i\ell}\partial_{x^\ell}$. Thus

$$
[X_i, X_j] = \sum_{\ell > k} \{X_i(a^{j\ell}) - X_j(a^{i\ell})\}\partial_{x^\ell}.
$$

Since $[X_i, X_j] \in C^\infty(V)$, $[X_i, X_j] = 0$. The theorem now follows from Lemma 2.7. $\square$

2.2.5 SUBMERSIONS. Let F be a smooth surjective map from a manifold M to a manifold $\tilde{M}$. We say that F is a *submersion* if $F_* : T_P M \to T_{F(P)}\tilde{M}$ is surjective for every point P of M. By the Implicit Function Theorem, the *fiber* $F_x := \pi^{-1}(\tilde{P})$ is a smooth submanifold of M for every point $\tilde{P} \in \tilde{M}$. We have:

Theorem 2.9   *Let $\pi : M \to \tilde{M}$ be a smooth surjective map from M to $\tilde{M}$.*

1. *If $\pi : M \to \tilde{M}$ is a fiber bundle, then π is a submersion.*

2. *If $\pi : M \to \tilde{M}$ is a submersion, if M is compact, and if $\tilde{M}$ is connected then $(M, \pi, \tilde{M})$ is a fiber bundle.*

Proof. Assertion 1 is immediate from the definition. Assume π is a submersion and M is compact; this implies $\tilde{M}$ is compact. Use a partition of unity to put a Riemannian metric on M, i.e., a positive definite inner product on the tangent bundle TM. By assumption, the map $\pi_* : T_P M \to T_{\pi(P)}\tilde{M}$ is surjective. We may split $TM = \mathcal{H} \oplus \mathcal{V}$ where the vertical space $\mathcal{V} = \ker\{\pi_*\}$ is the tangent space of the fibers and $\mathcal{H} = \mathcal{V}^\perp$. We then have π_* is an isomorphism from $\mathcal{H}_P$ to $T_{\pi(P)}\tilde{M}$. Thus, if $\tilde{X}$ is a smooth vector field on $\tilde{M}$, we can lift $\tilde{X}$ to a unique vector field X on M by requiring that $X \in \mathcal{H}$ and $\pi_*(X) = \tilde{X}$. If $\gamma(t)$ is an integral curve for X in M, then $\pi(\gamma(t))$ is an integral curve for $\tilde{X}$ in $\tilde{M}$. Let $\tilde{P} \in \tilde{M}$ and let $(\tilde{\mathcal{O}}, (\tilde{x}^1, \ldots, \tilde{x}^m))$ be local coordinates centered at $\tilde{P}$. Let $\tilde{X}_i := \partial_{\tilde{x}^i}$ be vector fields on $\tilde{\mathcal{O}}$ and let X_i be the lifted vector fields on $\mathcal{O} = \pi^{-1}(\tilde{\mathcal{O}})$. If $\vec{t} \in \mathbb{R}^m$, let $X_{\vec{t}}$ be a coordinate vector field, i.e.,

$$X_{\vec{t}} := t^1 X_1 + \cdots + t^m X_m \quad \text{and} \quad \tilde{X}_{\vec{t}} := \pi_*(X_{\vec{t}}) = t^1 \tilde{X}_1 + \cdots + t^m \tilde{X}_m.$$

Let $\Phi(s, \vec{t}, y)$ be the integral curve for $X_{\vec{t}}$ starting at a point y in the fiber over $\tilde{P}$. Since the fibers are compact, the Fundamental Theorem of Ordinary Differential Equations shows that there exists $\delta > 0$ so these curves extend for $0 \le s \le 1$ if $\|\vec{t}\| \le \delta$. Let $\Phi_1(\vec{t}, y) := \Phi(1, \vec{t}, y)$. Because we have that $\pi_*(X_{\vec{t}}) = \tilde{X}_{\vec{t}}$, $\pi(\Phi(s, \vec{t}, y)) = s\vec{t}$ and, consequently, $\pi(\Phi_1(\vec{t}, y)) = \vec{t}$. This constructs a diffeomorphism from $B_\delta(\tilde{P}) \times F$ to $\pi^{-1}(B_\delta(\tilde{P}))$ and shows π is a fiber bundle. It shows all the fibers over $B_\delta(\tilde{P})$ are diffeomorphic and thus, as $\tilde{M}$ is connected, all the fibers are diffeomorphic. □

2.2.6 THE COTANGENT BUNDLE. The *cotangent bundle* T^*M is the dual of the tangent bundle. Let $\ll \cdot, \cdot \gg$ denote the natural paring between these bundles. Let $(\mathcal{O}, \phi)$ be local coordinates where $\phi = (x^1, \ldots, x^m)$. If $f \in C^\infty(M)$, then the *exterior derivative* of f at P is the element of $T_P^*(M)$ characterized by the identity

$$\ll df, X \gg = X(f)(P) \quad \text{for all} \quad X \in T_P M \quad \text{or equivalently} \quad df = \partial_{x^i} f \cdot dx^i.$$

A smooth (local) section to T^*M is called a 1-form; the collection $\{dx^1, \ldots, dx^m\}$ is a local frame for $T^*\mathcal{O}$ which is dual to the coordinate frame $\{\partial_{x^1}, \ldots, \partial_{x^m}\}$ for the tangent bundle $T\mathcal{O}$. If $(y^1, \ldots, y^m)$ is another set of coordinates, then Equation (2.2.a) dualizes to become:

$$dy^i = \frac{\partial y^i}{\partial x^j} \cdot dx^j.$$

If F is a smooth map from M to N, then Equation (2.2.c) dualizes to give a *pullback map* F^* from $T_{F(P)}^* N$ to $T_P^* M$ so that

$$\ll \omega, F_* X \gg = \ll F^* \omega, X \gg \quad \text{for} \quad X \in T_P M \quad \text{and} \quad \omega \in T_{F(P)}^*(N), \tag{2.2.d}$$

or, equivalently, $F^*(dy^i) = \frac{\partial y^i}{\partial x^j} dx^j$. We have $\text{Id}^* = \text{Id}$ and $(F \circ G)^* = G^* \circ F^*$. The Chain Rule yields the intertwining formula:

$$d_M(F^* f) = F^* d_N(f).$$

2.2.7 THE GEOMETRY OF THE COTANGENT BUNDLE. The cotangent bundle has, in many respects, a richer geometry than does the tangent bundle. We refer to Jost [22] and Yano and Ishihara [42] for further details concerning this material. Let $\sigma : T^*M \to M$ be the *natural* *projection* from the cotangent bundle to the base manifold. We may express any point $\tilde{P}$ of T^*M in the form $\tilde{P} = (P, \omega)$ where $P := \sigma(\tilde{P})$ belongs to M and where ω belongs to T_P^*M. Let $(\mathcal{O}, \phi)$ be a coordinate chart on M. Expanding $\omega = x_i\prime dx^i$ defines a system of local coordinates $(x^1, \ldots, x^m; x_{1\prime}, \ldots, x_{m\prime})$ on $\tilde{\mathcal{O}} := \sigma^{-1}(\mathcal{O}) \subset T^*M$. The dual coordinates are written with the index down as they transform covariantly rather than contravariantly; this permits us to retain the formalism of summing over repeated indices.

Let $X \in C^\infty(TM)$ be a smooth vector field on M. The *evaluation map* ιX is a smooth function on the cotangent bundle T^*M which is defined by the identity:

$$\iota X(P, \omega) = \omega(X_P).$$

We may express $X = X^i \partial_{x^i}$ to define the coefficients $X^i = \ll X, dx^i \gg$. We then have that

$$\iota X(x^i, x_{i\prime}) = x_{i\prime} X^i = x_{i\prime} \ll X, dx^i \gg .$$

Vector fields on T^*M are characterized by their action on the evaluation maps ιX (see Yano and Ishihara [42]):

Lemma 2.10 Suppose that $\tilde{Y}$ and $\tilde{Z}$ are smooth vector fields on the cotangent bundle T^*M. Suppose that $\tilde{Y}(\iota X) = \tilde{Z}(\iota X)$ for all smooth vector fields X on M. Then $\tilde{Y} = \tilde{Z}$.

Proof. Let $\tilde{Y} = a^i(\vec{x}, \vec{x}\prime)\partial_{x^i} + b_{i\prime}(\vec{x}, \vec{x}\prime)\partial_{x_{i\prime}}$ be a smooth vector field on T^*M with $\tilde{Y}(\iota X) = 0$ for all smooth vector fields on M. We must show this implies that $\tilde{Y} = 0$. Let $X = X^j(\vec{x})\partial_{x^j}$. Since $\iota X = x_{i\prime} X^i$, we have:

$$0 = \tilde{Y}(\iota X) = x_{j\prime} a^i(\vec{x}, \vec{x}\prime)(\partial_{x^i} X^j)(\vec{x}) + b_{i\prime}(\vec{x}, \vec{x}\prime) X^i(\vec{x}).$$

Fix j and do not sum. If we take $X = \partial_{x^j}$, then $X^i = \delta_j^i$ so $\iota X = x_{j\prime}$ and

$$0 = \tilde{Y}(\iota X) = b_{j\prime}(\vec{x}, \vec{x}\prime) \quad \text{so} \quad b_{j\prime} = 0 \quad \text{and} \quad \tilde{Y} = a^i(\vec{x}, \vec{x}\prime)\partial_{x^i} .$$

If we take $X = x^j \partial_{x^j}$, then we have similarly $0 = \tilde{Y}(\iota X) = x_{j\prime} a^j(\vec{x}, \vec{x}\prime)$. Thus $a^j(\vec{x}, \vec{x}\prime) = 0$ when $x_{j\prime} \neq 0$. Since the functions $a^j(\vec{x}, \vec{x}\prime)$ are smooth, this implies a^j vanishes identically and hence $\tilde{Y} = 0$ as desired. $\square$

Let X be a smooth vector field on M. The *complete lift* X^C is the vector field on the cotangent bundle T^*M characterized via Lemma 2.10 by the identity:

$$X^C(\iota Z) = \iota[X, Z] \quad \text{for all smooth vector fields } Z \text{ on } M.$$

Lemma 2.11 Let $\omega \in T_P^*M$ with $\omega \neq 0$. Then the tangent space $T_{(P,\omega)}T^*M$ is spanned by the complete lifts of all the smooth vector fields on M.

Proof. We first compute the complete lift in a system of local coordinates. Let

$$X = X^j(\vec{x})\partial_{x^j} \quad \text{and} \quad X^C = a^j(\vec{x}, \vec{x}')\partial_{x^j} + b_{j'}(\vec{x}, \vec{x}')\partial_{x_{j'}}.$$

Let $Z = Z^i(\vec{x})\partial_{x^i}$. Then

$$
\begin{aligned}
X^C(\iota Z) &= X^C(x_{i'}Z^i) = x_{i'}a^j(\vec{x}, \vec{x}')(\partial_{x^j}Z^i)(\vec{x}) + b_{j'}(\vec{x}, \vec{x}')Z^j(\vec{x}) \\
&= \iota\{(X^j\partial_{x^j}Z^i - Z^j\partial_{x^j}X^i)\partial_{x^i}\} \\
&= x_{i'}(X^j(\vec{x})(\partial_{x^j}Z^i)(\vec{x}) - Z^j(\vec{x})(\partial_{x^j}X^i)(\vec{x})).
\end{aligned}
$$

Since Z is arbitrary, we conclude $a^j(\vec{x}, \vec{x}') = X^j(\vec{x})$ and $b_{j'}(\vec{x}, \vec{x}') = -x_{i'}\partial_{x^j}X^i$ so that

$$X^C = X^j(\vec{x})\partial_{x^j} - x_{i'}(\partial_{x^j}X^i)(\vec{x})\partial_{x_{j'}}.$$

Taking $X = \partial_{x^j}$ then yields $X^C = \partial_{x^j}$. Since $\omega \neq 0$, we have $x_{i'} \neq 0$ for some i. Furthermore, taking $X = x^j\partial_{x^i}$ then yields $X^C = x^j\partial_{x^i} - x_{i'}\partial_{x_{j'}}$ which completes the proof. We note that this fails on the 0-section. Since the zero section is defined by the relation $x_{i'} = 0$, $=X^C = X^j(\vec{x})\partial_{x^j}$ does not involve $\partial_{x_{j'}}$. Consequently on the zero section,

$$\text{span}\{X^C\} = \text{span}\{\partial_{x^j}\}. \qquad \square$$

A crucial point is that since ιX and X^C are invariantly characterized, we do not need to check the local formalism transforms correctly; on the other hand, the local formalism shows that X^C in fact exists. We shall apply similar arguments subsequently.

Lemma 2.12 Two smooth tensor fields Ψ_1, Ψ_2 of type $(0, s)$ on T^*M coincide with each other if and only if we have the following identity for all vector fields X_i on M:

$$\Psi_1(X_{i_1}^C, \ldots, X_{i_s}^C) = \Psi_2(X_{i_1}^C, \ldots, X_{i_s}^C).$$

Proof. The Lemma is immediate if $\omega \neq 0$ by Lemma 2.11; we now use continuity to extend the result to the 0-section. $\qquad \square$

Let $T \in C^\infty(\text{Hom}(TM))$ be a smooth $(1, 1)$-tensor field on M. Define a corresponding lifted 1-form $\iota T \in C^\infty(T^*(T^*M))$ on the cotangent bundle characterized by:

$$\ll \iota T, X^C \gg = \iota(TX).$$

We compute in local coordinates as follows. Expand $\iota T = a_i(\vec{x}, \vec{x}')dx^i + b^i(\vec{x}, \vec{x}')dx_{i'}$ and $T\partial_{x^i} = T_i^j\partial_{x^j}$. Let $X = X^i\partial_{x^i}$. We compute:

$$
\begin{aligned}
(\iota T)(X^C) &= a_i(\vec{x}, \vec{x}')X^i(\vec{x}) - b^j(\vec{x}, \vec{x}')x_{i'}(\partial_{x^j}X^i), \\
\iota(TX) &= \iota(X^i T_i^j\partial_{x^j}) = x_{j'}X^i T_i^j.
\end{aligned}
$$

This shows that $b^j = 0$ and that $a_i = x_{j'}T_i^j$, i.e., that $\iota T = x_{j'}T_i^j dx^i$ where $T = T_i^j\partial_{x^j} \otimes dx^i$. We also refer to Kowalski and Sekizawa [25] for additional information concerning natural lifts in Riemannian geometry. The following result summarizes the formalisms which we have established:

Lemma 2.13

1. The evaluation map ιX for $X \in C^\infty(TM)$:

   (a) Invariant formalism: $\iota X(P, \omega) = \omega(X_P)$.

   (b) Coordinate formalism: $\iota X = x_{i'}X^i$ for $X = X^i \partial_{x^i}$.

2. The complete lift of a vector field $X \in C^\infty(TM)$.

   (a) Invariant formalism: $X^C(\iota Z) = \iota[X, Z]$ for $Z \in C^\infty(TM)$.

   (b) Coordinate formalism: $X^C = X^j \partial_{x^j} - x_{i'}(\partial_{x^j} X^i)\partial_{x_{j'}}$, for $X = X^j \partial_{x^j}$.

3. Let $T \in C^\infty(\mathrm{Hom}(TM))$; ιT is the 1-form on T^*M:

   (a) Invariant formalism: $\ll \iota T, X^C \gg = \iota(TX)$.

   (b) Coordinate formalism: $\iota T = x_{j'}T_i^j dx^i$.

2.3   STOKES' THEOREM

This section is devoted to the proof of the generalized Stokes' Theorem.

2.3.1 THE EXTERIOR ALGEBRA. Let V^* be the dual space of an r-dimensional real vector space V. The *exterior algebra* $(\Lambda(V^*), \wedge)$ is the universal real unital algebra generated by V subject to the relations $v \wedge w + w \wedge v = 0$. More formally, let

$$\mathcal{T} := \mathbb{R} \oplus V^* \oplus (V^* \otimes V^*) \oplus \cdots = \oplus_{n \geq 0} \{\otimes^n V^*\}$$

be the complete tensor algebra. Let $\mathcal{I}$ be the 2-sided ideal of $\mathcal{T}$ generated by all tensors of the form $v^1 \otimes v^2 + v^2 \otimes v^1$ for $v^i \in V^*$. Set $\Lambda(V^*) = \mathcal{T}/\mathcal{I}$. The ideal $\mathcal{I}$ is homogeneous; let

$$\mathcal{I}^k := \mathrm{span}\{v^{i_1} \otimes \cdots \otimes v^{i_\ell} \otimes (v^{i_{\ell+1}} \otimes v^{i_{\ell+2}} - v^{i_{\ell+2}} \otimes v^{i_{\ell+1}}) \otimes v^{i_{\ell+3}} \otimes \cdots \otimes v^{i_k}\} \quad (2.3.a)$$

to decompose $\Lambda(V^*) = \oplus_{k \geq 0} \Lambda^k(V^*)$ where $\Lambda^k(V^*) := \otimes^k V^*/\mathcal{I}^k$. Note that:

$$\mathcal{I}^0 = \{0\} \quad \text{and} \quad \mathcal{I}^1 = \{0\} \quad \text{so} \quad \Lambda^0(V^*) = \mathbb{R} \quad \text{and} \quad \Lambda^1(V^*) = V^*.$$

Let $\{e_1, \ldots, e_r\}$ be a basis for V and let $\{e^1, \ldots, e^r\}$ be the associated dual basis for V^*. Since $\{e^i\}$ spans V^*, $\{e^{i_1} \wedge \cdots \wedge e^{i_k}\}$ spans $\Lambda^k(V^*)$ as a vector space over $\mathbb{R}$ where $\{i_1, \ldots, i_k\}$ is an arbitrary collection of k indices between 1 and r. The defining relation shows $e^i \wedge e^j = -e^j \wedge e^i$ and hence we have $e^i \wedge e^i = 0$. If $I = (i_1, \ldots, i_k)$ is a collection of strictly increasing indices $1 \leq i_1 < \cdots < i_k \leq r$, set $e^I := e^{i_1} \wedge \cdots \wedge e^{i_k}$. Then

$$\Lambda^k(V^*) = \mathrm{span}\{e^I\}_{|I|=k}.$$

This shows $\dim(\Lambda^k(V^*)) \leq \binom{r}{k}$ if $0 \leq k \leq r$ and that $\Lambda^k(V^*) = \{0\}$ if $k > r$. If $|I| = k$ and $|J| = \ell$, we must pass k indices over ℓ indices to compare $e^I \wedge e^J$ with $e^J \wedge e^I$. Consequently

$$\omega_k \wedge \tilde{\omega}_\ell = (-1)^{k\ell} \tilde{\omega}_\ell \wedge \omega_k \quad \text{if} \quad \omega_k \in \Lambda^k(V^*) \quad \text{and} \quad \tilde{\omega}_\ell \in \Lambda^\ell(V^*).$$

There is a natural evaluation of $\Lambda^k(V^*)$ on $\times^k V$ that will play a central role in our subsequent discussion. Let $\vec{v} := (v_1, \ldots, v_k) \in \times^k V$ and let $\vec{v}^* := (v^1, \ldots, v^k) \in \times^k V^*$. Then $\ll v_i, v^j \gg_{1 \leq i,j \leq k}$ is a $k \times k$ real matrix. Let $\Theta_{\vec{v}}(\vec{v}^*) := \det(\ll v_i, v^j \gg) \in \mathbb{R}$. The map from $\vec{v}^*$ to $\Theta_{\vec{v}}(\vec{v}^*)$ is multilinear and, consequently, extends to a map $\Theta_{\vec{v}} : \otimes^k V^* \to \mathbb{R}$. Let π_Λ^k be the natural projection from $\otimes^k V^*$ to $\Lambda^k(V^*) = \otimes^k V^*/\mathcal{I}^k$. Since the determinant changes sign if we interchange two adjacent columns, Θ vanishes on the elements of Equation (2.3.a) and extends to a map $\Theta : \Lambda^k(V^*) \to \mathbb{R}$. If $\omega \in \Lambda^k(V^*)$, we define $\omega[v_1, \ldots, v_k] := \Theta_{\vec{v}}(\omega)$. We then have a well-defined evaluation (which is linear in ω) that we use subsequently:

$$(v^1 \wedge \cdots \wedge v^k)[v_1, \ldots, v_k] = \det(\ll v_i, v^j \gg). \tag{2.3.b}$$

There is always the danger when constructing an algebra using generators and relations that everything collapses; that there are somehow unforeseen relations that make everything trivial. Fortunately that is not the case in the setting at hand.

Lemma 2.14   Adopt the notation established above. Then $\{e^I\}_{|I|=k}$ is a basis for $\Lambda^k(V^*)$. Consequently, $\dim(\Lambda^k(V^*)) = \binom{r}{k}$ if $0 \leq k \leq r$ and $\Lambda^k(V^*) = \{0\}$ if $k > r$.

Proof. We have already noted $\Lambda^k(V^*) = \{0\}$ if $k > r$. If $k \leq r$, then the elements $\{e^I\}_{|I|=k}$ form a spanning set for $\Lambda^k(V^*)$. Let $J = (j_1, \ldots, j_r)$ for $1 \leq j_1 < \cdots < j_k \leq r$. We use the defining relation of Equation (2.3.b) to see that

$$e^I[e_{j_1}, \ldots, e_{j_k}] = \det(\ll e_{j_\mu}, e^{i_\nu} \gg_{\nu,\mu}) = \left\{ \begin{array}{ll} 0 & \text{if } I \neq J \\ 1 & \text{if } I = J \end{array} \right\}.$$

Since the pairing $\omega \to \omega[e_J]$ is a linear map from $\Lambda^k(V^*)$ to $\mathbb{R}$, this shows the $\{e^I\}_{|I|=k}$ form a basis for $\Lambda^k(V^*)$. □

The pairing $\omega[\cdot]$ described in Equation (2.3.b) identifies $\Lambda^k(V^*)$ with the set of multilinear totally alternating bilinear forms on V and embeds $\Lambda^k(V^*)$ as a subspace of $\otimes^k V^*$. Let $\Pi(k)$ be the set of all permutations of k elements. Let

$$\sigma_\Lambda^k(v^1 \otimes \cdots \otimes v^k) := \frac{1}{k!} \sum_{\rho \in \Pi(k)} v^{\rho(1)} \otimes \cdots \otimes v^{\rho(k)}.$$

Then $(\sigma_\Lambda^k)^2 = \sigma_\Lambda^k$ and $\text{range}\{\sigma_\Lambda^k\} = \Lambda^k(V^*)$ so σ_Λ^k provides a projection of $\otimes^k V^*$ onto $\Lambda^k(V^*)$. We add a note of caution. With our notational conventions, $\pi_\Lambda^k = k! \sigma_\Lambda^k$.

2.3.2 THE PULLBACK. If Ψ belongs to $\mathrm{End}(V, W)$, then the dual map Ψ^* belongs to $\mathrm{End}(W^*, V^*)$ and extends to a map $\mathcal{T}(\Psi^*)$ from $\mathcal{T}(W^*)$ to $\mathcal{T}(V^*)$. By Equation (2.3.a), $\mathcal{T}(\Psi^*)$ maps the defining ideal for W to the defining ideal for V. Consequently, we have naturally defined linear maps $\Lambda^k(\Psi^*)$ from $\Lambda^k(W^*)$ to $\Lambda^k(V^*)$. Let $\{e^i\}$ be a basis for V^*, let $\{f^a\}$ be a basis for W^*, and let $\Psi^*(f^a) = \Psi_i^a e^i$. Let $e^{ij} := e^i \wedge e^j$ for $i < j$ and $e^{ijk} := e^i \wedge e^j \wedge e^k$ for $i < j < k$. We have, for example:

$$\Psi^*(f^a) = \Psi_i^a e^i, \qquad \Psi^*(f^a \wedge f^b) = \Psi_i^a \Psi_j^b e^i \wedge e^j = (\Psi_i^a \Psi_j^b - \Psi_j^a \Psi_i^b) e^{ij},$$
$$\Psi^*(f^a \wedge f^b \wedge f^c) = \Psi_i^a \Psi_j^b \Psi_k^c e^i \wedge e^j \wedge e^k$$
$$= (\Psi_i^a \Psi_j^b \Psi_k^c + \Psi_j^a \Psi_k^b \Psi_i^c + \Psi_k^a \Psi_i^b \Psi_j^c - \Psi_i^a \Psi_k^b \Psi_j^c - \Psi_k^a \Psi_j^b \Psi_i^c - \Psi_j^a \Psi_i^b \Psi_k^c) e^{ijk}.$$

The coefficients are, of course, just the determinant of the minors Ψ_μ^ν for $\mu \in \{i_1, \dots, i_k\}$ and $\nu \in \{a_1, \dots, a_k\}$. Thus, for example, if $V = W$ and if $k = r$, we have

$$\Psi^*(e^1 \wedge \cdots \wedge e^r) = \det(\Psi) e^1 \wedge \cdots \wedge e^r.$$

We have $(\Psi_1 \circ \Psi_2)^* = \Psi_2^* \circ \Psi_1^*$ and $\mathrm{Id}^* = \mathrm{Id}$. In the language of algebraic topology, the association $V \rightsquigarrow \Lambda^k(V)$ is a *contravariant functor* from the *category* of finite-dimensional vector spaces to the category of *graded skew-commutative unital rings*. It is not necessary to fuss unduly about this notation. We refer to Section 8.1.1 of Book II for further details concerning category theory.

Let V be a smooth vector bundle over a manifold M. The construction of $\Lambda^k(\cdot)$ is *natural*, i.e., it is well-defined and is independent of the basis chosen. Thus the vector spaces $\Lambda^k(V_x^*)$ patch together to define a smooth vector bundle $\Lambda^k(V^*)$ over M. If $\Psi_{\alpha\beta}^*$ are the transition functions of V^*, then $\Lambda^k(\Psi_{\alpha\beta}^*)$ are the transition functions of $\Lambda^k(V^*)$. Let V_i be vector bundles over M. If $\Psi : V_1 \to V_2$ is a bundle map, then $\Lambda^k(\Psi^*)$ is a bundle map from $\Lambda^k(V_2)$ to $\Lambda^k(V_1)$.

2.3.3 DIFFERENTIAL FORMS. We take V to be the cotangent bundle T^*M and simply the notation by setting $\Lambda^k M := \Lambda^k(T^*M)$. The space of *smooth k-forms* over M is the space of smooth sections to $\Lambda^k M$. Let $\omega \in C^\infty(\Lambda^k M)$. If $\vec{x} = (x^1, \dots, x^m)$ is a system of local coordinates on M, expand

$$\omega = a_I(x) dx^I \quad \text{where} \quad dx^I := dx^{i_1} \wedge \cdots \wedge dx^{i_k} \quad \text{and} \quad I = \{1 \le i_1 < \cdots < i_k \le m\}.$$

If $\vec{y} = (y^1, \dots, y^m)$ is another system of local coordinates on M, then on the common domain of definition, we have:

$$dy^I = \frac{\partial y^I}{\partial x^J} dx^J \quad \text{where} \quad \frac{\partial y^I}{\partial x^J} := \det\left(\frac{\partial y^i}{\partial x^j}\right)_{i \in I, j \in J}. \tag{2.3.c}$$

In particular, if $k = m$, then the transition function is the Jacobian determinant:

$$dy^1 \wedge \cdots \wedge dy^m = \det\left(\frac{\partial y^i}{\partial x^j}\right) dx^1 \wedge \cdots \wedge dx^m. \tag{2.3.d}$$

If F is a smooth map from a manifold M to a manifold N, then we have the pushforward map $F_* : T_P M \to T_{F(P)} N$ and the pullback map $F^* : T^*_{F(P)} N \to T^*_P M$. This extends to linear maps $F^* : \Lambda^k_{F(P)} N \to \Lambda^k_P M$. We have linear maps

$$F^* : C^\infty(\Lambda^k N) \to C^\infty(\Lambda^k M) \quad \text{so} \quad F^*(\omega_k \wedge \tilde{\omega}_\ell) = F^*(\omega_k) \wedge F^*(\tilde{\omega}_\ell).$$

If $(x^1, \ldots, x^m)$ are local coordinates on M and if $(y^1, \ldots, y^n)$ are local coordinates on N, then Equation (2.3.c) generalizes immediately to become:

$$F^*(a_I(y)dy^I) = a_I(F(x))\frac{\partial y^I}{\partial x^J}dx^J \quad \text{where} \quad \frac{\partial y^I}{\partial x^J} := \det\left(\frac{\partial F^* y^i}{\partial x^j}\right)_{i \in I, j \in J}.$$

If G is a smooth map from N to S, then we have that $(G \circ F)^* = F^* \circ G^*$; $\mathrm{Id}^* = \mathrm{Id}$.

Lemma 2.15   If f is a smooth function on M, let df be the smooth 1-form characterized by $\ll df, \xi \gg = \xi(f)$ for any smooth vector field ξ on M; if $\vec{x} = (x^1, \ldots, x^m)$ is a system of local coordinates on M, then $df = \partial_{x^i} f \cdot dx^i$.

1. There is a unique extension of d to map $C^\infty(\Lambda^k M)$ to $C^\infty(\Lambda^{k+1} M)$ for $k \geq 1$ satisfying $d^2 = 0$ and $d(\omega_p \wedge \tilde{\omega}_q) = d\omega_p \wedge \tilde{\omega}_q + (-1)^p \omega_p \wedge d\tilde{\omega}_q$ for $\omega_p \in C^\infty(\Lambda^p M)$ and $\tilde{\omega}_q \in C^\infty(\Lambda^q M)$.

2. $d(f_I dx^I) = df_I \cdot dx^I$.

3. If F is a smooth map from M to N, then $d_M F^* = F^* d_N$.

Proof. Since $\ll df, \xi \gg = \xi(f)$, $df = \partial_{x^i} f \cdot dx^i$. Suppose there exists an operator d which satisfies the properties of Assertion 1. We first establish that d is *local*. Suppose $\mathcal{O}$ is an open subset of M and that $\omega_1 = \omega_2$ on $\mathcal{O}$. Let $P \in \mathcal{O}$. Choose a smooth mesa function η so η is identically 1 near P and so η has support in $\mathcal{O}$. Since $\eta(\omega_1 - \omega_2)$ vanishes identically,

$$\begin{aligned} 0 &= d(\eta(\omega_1 - \omega_2))(P) = d\eta(P)(\omega_1(P) - \omega_2(P)) + \eta(P)(d\omega_1(P) - d\omega_2(P)) \\ &= 0 + d\omega_1(P) - d\omega_2(P). \end{aligned}$$

Thus $\omega_1 = \omega_2$ on $\mathcal{O}$ implies $d\omega_1 = d\omega_2$ on $\mathcal{O}$. Let $\vec{x} = (x^1, \ldots, x^m)$ be a system of coordinates defined on an open subset $\mathcal{O}$ of M. Fix $P \in \mathcal{O}$. Choose a mesa function η_1 which is identically 1 near P with support in $\mathcal{O}$. Choose a second mesa function η_2 which is identically 1 on the support of η_1. Then $d(\eta_2 x^i) = dx^i$ near P. We may then expand:

$$\begin{aligned} \eta_1 \omega &= \eta_1 f_{(i_1, \ldots, i_k)} d(\eta_2 x^{i_1}) \wedge \cdots \wedge d(\eta_2 x^{i_k}), \\ d(\eta_1 \omega) &= d(\eta_1 f_{(i_1, \ldots, i_k)}) d(\eta_2 x^{i_1}) \wedge \cdots \wedge d(\eta_2 x^{i_k}) \\ &+ \sum_{\nu=1}^k (-1)^\nu (\eta_1 f_{(i_1, \ldots, i_k)}) d(\eta_2 x^{i_1}) \wedge \cdots \wedge d\, d(\eta_2 x^{i_\nu}) \wedge \cdots \wedge d(\eta_2 x^{i_k}). \end{aligned}$$

Since $dd(\eta_2 x^{iv}) = 0$, since $d(\eta_1)(P) = 0$, and since $\eta_1(P) = 1$, we have:

$$d(\omega)(P) = d(\eta_1\omega)(P) = \{d(f_{(i_1,\ldots,i_k)})dx^{i_1} \wedge \cdots \wedge dx^{i_k}\}(P).$$

This establishes the formula of Assertion 2 and shows that if d exists, then it is unique.

To establish existence, we work in a coordinate chart. We use Assertion 2 to define d_x on $\mathcal{O}$. We show $d_x^2 = 0$ by computing:

$$d_x^2\left(f_I dx^I\right) = d_x\left(\partial_{x^i} f_I \cdot dx^i \wedge dx^I\right) = \partial_{x^i}\partial_{x^j} f \cdot dx^j \wedge dx^i \wedge dx^I.$$

This vanishes since $\partial_{x^i}\partial_{x^j} f = \partial_{x^j}\partial_{x^i} f$ but $dx^i \wedge dx^j = -dx^j \wedge dx^i$. If $f, g \in C^\infty(M)$, then $d_x(fg) = g d_x f + f d_x g$. We complete the proof of Assertion 1 by computing:

$$\begin{aligned}
d_x(\omega_p \wedge \tilde{\omega}_q) &= d_x\left(f_I dx^I \wedge g_J dx^J\right) = d_x(f_I g_J) \wedge dx^I \wedge dx^J \\
&= g_J d_x(f_I) \wedge dx^I \wedge dx^J + f_I d_x(g_J) \wedge dx^I \wedge dx^J \\
&= \left(d_x f \wedge dx^I\right) \wedge \left(g_J \wedge dx^J\right) + (-1)^p \left(f_I dx^I\right) \wedge \left(d_x g_J \wedge dx^J\right).
\end{aligned}$$

This shows $d := d_x$ exists on a coordinate chart. On the intersection of two coordinate charts, we must have $d_x = d_y$ by the uniqueness assertion. We can now define d globally with the desired properties using a partition of unity. This establishes Assertions 1 and 2.

Let $F : M \to N$. We complete the proof by computing:

$$\begin{aligned}
dF^*(dy^i) &= ddF^*(y^i) = 0, \\
dF^*(dy^I) &= \sum_\nu (-1)^{\nu-1} F^*(dy^{i_1}) \wedge \cdots \wedge dF^*(dy^{i_\nu}) \wedge \cdots \wedge F^*(dy^{i_k}) = 0, \\
d(F^*(f_I dy^I)) &= d(F^* f_I) \wedge F^*(dy^I) + F^*(f_I) \wedge dF^*(dy^I) \\
&= F^*(df_I) \wedge F^*(dy^I) = F^*(df_I \wedge dy^I) = F^* d(f_I dy^I).
\end{aligned}$$
$\qquad\qquad\qquad\qquad\qquad\qquad\qquad\qquad\qquad\qquad\qquad\qquad\qquad\qquad\qquad\square$

2.3.4 THE EXTERIOR DERIVATIVE AND THE LIE BRACKET.

There is a useful relationship between the Lie bracket and the exterior derivative that can be used to give an invariant definition of the action of d on 1-forms:

Theorem 2.16 *Let ω be a 1-form on a manifold M and let X and Y be vector fields on M. Then $d\omega(X, Y) = X(\omega(Y)) - Y(\omega(X)) - \omega([X, Y])$.*

Proof. Let $\vec{x} = (x^1, \ldots, x^m)$ be a system of local coordinates on M. Expand

$$\omega = a_i dx^i, \quad X = b^j \partial_{x^j}, \quad Z = c^k \partial_{x^k}.$$

Let $f_{/a} := \partial_{x^a} f$. We complete the proof by comparing the first line with the sum of the next three lines in the following display:

$$\begin{aligned}
d\omega([X, Y]) &= a_{i/\ell} dx^\ell \wedge dx^i (b^j \partial_{x^j}, c^k \partial_{x^k}) = a_{i/\ell}(b^\ell c^i - b^i c^\ell), \\
X(\omega(Y)) &= X(a_i c^i) = b^j a_{i/j} c^i + b^j a_i c^i_{/j}, \\
Y(\omega(X)) &= Y(a_i b^i) = c^k a_{i/k} b^i + c^k a_i b^i_{/k}, \\
\omega([X, Y]) &= \omega(b^j c^k_{/j} \partial_{x^k} - c^k b^j_{/k} \partial_{x^j}) = a_i b^j c^i_{/k} - a_i c^k b^i_{/k}.
\end{aligned}$$
$\qquad\qquad\qquad\qquad\qquad\qquad\qquad\qquad\qquad\qquad\qquad\qquad\qquad\qquad\qquad\square$

We note that the bracket can be used to give an invariant definition of the exterior derivative. We shall omit the verification as it plays no role in our subsequent development.

$$d\omega(X_0, X_1, \ldots, X_k) = \sum_{\nu}(-1)^{\nu} X_{\nu}(\omega(X_0, \ldots, \hat{X}_i, \ldots, X_k))$$
$$+ \sum_{\nu < \mu}(-1)^{\nu+\mu}\omega([V_{\nu}, V_{\mu}], X_0, \ldots, \hat{X}_i, \ldots, \hat{X}_j, \ldots, X_k).$$

2.3.5 ORIENTABILITY. M is said to be *oriented* by a coordinate atlas $\{(\mathcal{O}_{\alpha}, \phi_{\alpha})\}$ if the transition functions have positive Jacobian determinants, i.e., $\det(\phi'_{\alpha\beta} > 0)$. M is said to be *orientable* if it is oriented by some coordinate atlas. The Klein bottle is not orientable. On the other hand, the torus $\mathbb{T}^2 = S^1 \times S^1$ is orientable. If M_c is the level set of a smooth map f from $\mathbb{R}^{m+1}$ to $\mathbb{R}$ with df non-vanishing on M_c, then M_c is orientable. Thus the sphere S^m is orientable for any m.

Lemma 2.17 A connected manifold M of dimension m is orientable if and only if there exists a non-vanishing m-form on M, i.e., if the bundle $\Lambda^m M$ is trivial.

Proof. Suppose that there exists a non-vanishing m-form θ_m on M. Fix a coordinate atlas $\{(\mathcal{O}_{\alpha}, (x^1_{\alpha}, \ldots, x^m_{\alpha}))\}$ for M; we may assume without loss of generality that the $\mathcal{O}_{\alpha}$ are connected. Express $\theta_m = f_{\alpha} dx^1_{\alpha} \wedge \cdots \wedge dx^m_{\alpha}$ on $\mathcal{O}_{\alpha}$. Since θ_m is nowhere vanishing and $\mathcal{O}_{\alpha}$ is connected, f_{α} does not change sign. If $f_{\alpha} < 0$, we interchange the first two coordinates to create a new coordinate system where $f_{\alpha} > 0$. Thus we may assume $f_{\alpha} > 0$ for all α. Let $\Psi^{\beta}_{\alpha} = \det\left(\frac{\partial x^i_{\alpha}}{\partial x^j_{\beta}}\right)$; express

$$dx^1_{\alpha} \wedge \cdots \wedge dx^m_{\alpha} = \Psi^{\beta}_{\alpha} dx^1_{\beta} \wedge \cdots \wedge dx^m_{\beta} \qquad (2.3.e)$$

and hence $f_{\beta} = \Psi^{\beta}_{\alpha} f_{\alpha}$. This implies $\Psi^{\beta}_{\alpha} > 0$; we use θ_m to adjust the local coordinate systems appropriately and create an oriented atlas. Conversely, suppose M is orientable. Let $\{(\mathcal{O}_{\alpha}, (x^1_{\alpha}, \ldots, x^m_{\alpha}))\}$ be a coordinate atlas giving the orientation. Let ϕ^{α} be a partition of unity subordinate to the cover $\mathcal{O}_{\alpha}$ of M. Let $\theta_m := \sum_{\alpha} \phi^{\alpha}(dx^1_{\alpha} \wedge \cdots \wedge dx^m_{\alpha})$ on M; θ_m is well defined since $\phi^{\alpha} = 0$ on $\mathcal{O}^c_{\alpha}$ where $dx^1_{\alpha} \wedge \cdots \wedge dx^m_{\alpha}$ is not defined. Fix β. We use Equation (2.3.e) to express θ_m in terms of $dx^1_{\beta} \wedge \cdots \wedge dx^m_{\beta}$ on $\mathcal{O}_{\beta}$. The convex combination of positive functions is positive. Consequently $\sum_{\alpha} \phi^{\alpha}\Psi^{\beta}_{\alpha} > 0$ and

$$\theta_m = \left\{\sum_{\alpha} \phi^{\alpha}\Psi^{\beta}_{\alpha}\right\} dx^1_{\beta} \wedge \cdots \wedge dx^m_{\beta} \neq 0 \text{ on } \mathcal{O}_{\beta}. \qquad \square$$

2.3.6 PROJECTIVE SPACE. We continue the discussion of Section 2.1.6. Let $\mathbb{F}$ be the field of real numbers $\mathbb{R}$, the field of complex numbers $\mathbb{C}$, or the skew-field of quaternion numbers $\mathbb{H}$ as discussed in Section 1.2.1. Let $\mathbb{F}^* := \mathbb{F} - \{0\}$ be the group (with respect to multiplication) of non-zero elements of $\mathbb{F}$. Let $\mathbb{F}^*$ act on $\mathbb{F}^{m+1} - \{0\}$ by scalar multiplication from the left and let

$$\mathbb{F}\mathbb{P}^m := \{\mathbb{F}^{m+1} - \{0\}\}/\mathbb{F}^*$$

be the quotient space; this is the set of all $\mathbb{F}$-lines through the origin in $\mathbb{F}^{m+1}$. Let π be the natural projection from $\mathbb{F}^{m+1} - \{0\}$ to $\mathbb{F}\mathbb{P}^m$. We give $\mathbb{F}\mathbb{P}^m$ the *quotient topology*; a set $\mathcal{O}$ in $\mathbb{F}\mathbb{P}^m$ is open if and only if $\pi^{-1}(\mathcal{O})$ is open in $\mathbb{F}^{m+1} - \{0\}$; π is then an open map. Let X be a topological space. A map $\tilde{f} : \mathbb{F}\mathbb{P}^m \to X$ is continuous if and only if $\tilde{f} \circ \pi$ is continuous. Correspondingly, if $f : \mathbb{F}^{m+1} - \{0\} \to X$ is continuous and if $f(\lambda x) = f(x)$ for all $\lambda \in \mathbb{F}^*$ and all $x \in \mathbb{F}^{m+1} - \{0\}$, then f descends to a continuous map $\tilde{f} : \mathbb{F}\mathbb{P}^m \to X$. Let S be the unit sphere in $\mathbb{F}^{m+1}$. Since S is compact and $\pi : S \to \mathbb{F}\mathbb{P}^m$ is surjective, $\mathbb{F}\mathbb{P}^m$ is compact.

Let $\bar{x}$ be the conjugation operator of Section 1.2.1; $\bar{x}x = x\bar{x} = \|x\|^2$ and $\overline{uv} = \bar{v}\bar{u}$; the order is important for the quaternions as quaternionic multiplication is not commutative. The map $x \to \bar{x}$ is $\mathbb{R}$ linear. Let $(x^1, \ldots, x^{m+1}) \in \mathbb{F}^{m+1} - \{0\}$. We define the matrix $T(x) \in M_{m+1}(\mathbb{F})$ by setting

$$T_{ij}(x) := \|x\|^{-2}\bar{x}^i x^j \,.$$

Since $T_{ij}(\lambda x) = \|\lambda x\|^{-2}\bar{x}^i \bar{\lambda}\lambda x^j = |\lambda|^{-2}|\lambda|^2 \|x\|^{-2}\bar{x}^i x^j = T_{ij}(x)$, T extends to a continuous map from $\mathbb{F}\mathbb{P}^m$ to $M_{m+1}(\mathbb{F})$. Let $x \in S$ and let $\tilde{T}(x)$ be orthogonal projection on the line through x. Then $\tilde{T}(x)y = (y, x) \cdot x$ where $(y, x) = y^1\bar{x}^1 + \cdots + y^{m+1}\bar{x}^{m+1}$. Thus

$$(\tilde{T}(x)e_i, e_j) = (e_i, x)(x, e_j) = \bar{x}^i x^j \quad \text{so} \quad T = \tilde{T} \,.$$

Thus if $\sigma \in \mathbb{F}\mathbb{P}^m$, then range$\{T\sigma\} = \sigma$. This shows that T is an injective map from $\mathbb{F}\mathbb{P}^m$ to $M_{m+1}(\mathbb{F})$. Since $\mathbb{F}\mathbb{P}^m$ is compact, T is a homeomorphism to its image; we may identify $\mathbb{F}\mathbb{P}^m$ with the orthogonal projections of rank 1 in $M_{m+1}(\mathbb{F})$.

We give $\mathbb{F}\mathbb{P}^m$ the structure of a smooth manifold as follows. Let

$$U_i := \{x \in \mathbb{F}^{m+1} - \{0\} : x^i \neq 0\}$$

be an open cover of $\mathbb{F}^{m+1} - \{0\}$ for $1 \leq i \leq m + 1$. Since $\mathbb{F}^* \cdot U_i = U_i$, $\mathcal{O}_i := \pi(U_i)$ is an open cover of $\mathbb{F}\mathbb{P}^m$. Let $\Phi_i(x) := (x_i^{-1}x^1, \ldots, x_i^{-1}x^{m+1})$ on U_i. Note that $\Phi_i(x_1) = \Phi_i(x_2)$ if and only $x_1 = \lambda x_2$ and thus Φ_i descends to a continuous injective map from $\mathcal{O}_i$ to $\mathbb{F}^m$ where we omit the ith coordinate since it is identically 1. We set $y_i^k := (x^i)^{-1}x^k$ to obtain coordinates on $\mathcal{O}_i$. On $\mathcal{O}_i \cap \mathcal{O}_j$, we have

$$y_i^k = (x^i)^{-1}x^j(x^j)^{-1}x^k = \{(x^j)^{-1}x^i\}^{-1}(x^j)^{-1}x^k = (y_j^i)^{-1}y_j^k \,.$$

Thus the transition functions are smooth and give $\mathbb{F}\mathbb{P}^m$ the structure of a smooth manifold. We have that:

$$
\begin{aligned}
T_{ab} &= \|x\|^{-2}\bar{x}^a x^b = |x^i|^2\|x\|^{-2}\bar{x}^a(\bar{x}^i)^{-1}(x^i)^{-1}x^b = \{|x^i|^{-2}\|x\|^2\}^{-1}\bar{y}_i^a y_i^b \\
&= (1 + |y_i|^2)^{-1}\bar{y}_i^a y_i^b \,.
\end{aligned}
$$

This is the quotient of two polynomials in the real and imaginary parts of y and hence is a smooth function of $y_i = (y_i^1, \ldots, y_i^m)$. It is not difficult to verify that the Jacobian of T is injective so that $\mathbb{F}\mathbb{P}^m$ is a smooth submanifold of $M_{m+1}(\mathbb{F})$. If $\mathbb{F} = \mathbb{R}$, the coordinate systems defined above are real analytic; if $\mathbb{F} = \mathbb{C}$, the coordinate systems defined above are holomorphic; we postpone a further discussion of complex projective space until Section 4.3.3 in Book II. We now turn to the case of real projective space.

Lemma 2.18 $\mathbb{RP}^m$ is orientable if and only if m is odd.

Proof. Let $\omega_i = dy_i^1 \wedge \cdots \wedge dy_i^{i-1} \wedge dy_i^{i+1} \wedge \cdots \wedge dy_i^{m+1}$ be a non-zero m-form on $\mathcal{O}_i$. We examine $\mathcal{O}_i \cap \mathcal{O}_j$. We suppose $i < j$. We have: $y_i^k = (y_j^i)^{-1} y_j^k$. Consequently

$$
dy_i^k = \left\{
\begin{array}{ll}
(y_j^i)^{-1} dy_j^k - (y_j^i)^{-2} y_j^k dy_j^i & \text{if } k \neq i, j \\
-(y_j^i)^{-2} dy_j^i & \text{if } k = j \\
0 & \text{if } k = i
\end{array}
\right\} .
$$

Consequently,

$$
\omega_i = -(y_j^i)^{-m-1} dy_j^1 \wedge \cdots \wedge dy_j^{i-1} \wedge dy_j^{i+1} \wedge \cdots \wedge dy_j^{j-1} \wedge dy_j^i \wedge dy_j^{j+1} \wedge \cdots \wedge dy_j^{m+1}
$$

$$
= -(y_j^i)^{-m-1}(-1)^{i+j-1} dy_j^1 \wedge \cdots \wedge dy_j^{i-1} \wedge dy_j^i \wedge \cdots \wedge dy_j^{j-1}
$$

$$
\wedge dy_j^{j+1} \wedge \cdots \wedge dy_j^{m+1}
$$

$$
= (-1)^{i+j}(y_j^i)^{-m-1} \omega_j .
$$

Suppose m is odd. We let $\theta_i := (-1)^i \omega_i$. We then have:

$$
\theta_i = (-1)^i \omega_i = (-1)^i (-1)^{i+j} (y_j^i)^{-m-1} \omega_j = (y_j^i)^{-m-1} \theta_j .
$$

Since $m + 1$ is even, $(y_j^i)^{-m-1} > 0$ and $\mathbb{RP}^m$ is orientable. If m is even, then $(y_j^i)^{-m-1}$ changes sign on $\mathcal{O}_i \cap \mathcal{O}_j$ and thus there is no possibility of arranging the orientations to agree. Thus $\mathbb{RP}^m$ is not orientable if m is even. $\square$

2.3.7 MANIFOLDS WITH BOUNDARY.

We consider the *lower half space*

$$
\mathbb{R}_-^m := \{x = (x^1, \ldots, x^m) \in \mathbb{R}^m : x^1 \leq 0\} .
$$

The boundary of $\mathbb{R}_-^m$ is the hyperplane $\mathbb{R}^{m-1} := \{x \in \mathbb{R}_-^m : x^1 = 0\}$. Let $\mathcal{O}$ be an open subset of $\mathbb{R}_-^m$ regarded as a metric space in its own right; $\mathcal{O}$ need not be an open subset of $\mathbb{R}^m$. We say that F is smooth on $\mathcal{O}$ if there is an open subset $\tilde{\mathcal{O}}$ of $\mathbb{R}^m$ which contains $\mathcal{O}$ and a smooth function $\tilde{F}$ on $\tilde{\mathcal{O}}$ such that F is the restriction of $\tilde{F}$ to $\mathcal{O}$. Let $\mathrm{bd}(M)$ be a closed subspace of a metric space M. Let $\{\mathcal{O}_\alpha\}$ be an open cover of M. If $\mathcal{O}_\alpha \cap \mathrm{bd}(M)$ is empty, we assume given homeomorphisms from $\mathcal{O}_\alpha$ to an open subset $\mathcal{U}_\alpha$ of $\mathbb{R}^m$. If $\mathcal{O}_\alpha \cap \mathrm{bd}(M)$ is non-empty, we assume given homeomorphisms from $\mathcal{O}_\alpha$ to open subsets $\mathcal{U}_\alpha$ of $\mathbb{R}_-^m$ such that $\phi_\alpha(\mathcal{O}_\alpha \cap \mathrm{bd}(M)) \subset \mathbb{R}^{m-1}$. We assume that the transition functions $\phi_{\alpha\beta} := \phi_\alpha \circ \phi_\beta^{-1}$ are smooth on the appropriate domain. In this setting, the restriction of the coordinate charts to $\mathrm{bd}(M)$ gives $\mathrm{bd}(M)$ the structure of a smooth embedded submanifold of M and $M - \text{bd}(M)$ is a smooth open manifold. The pair $(M, \text{bd}(M))$ is said to be a *manifold with boundary*.

The unit sphere S^{m-1} is the boundary of the unit disk D^m. Every manifold N is the boundary of the non-compact manifold $(-\infty, 0] \times N$. However, there are compact manifolds N which are not the boundary of any compact manifold. For example, complex projective space $\mathbb{CP}^2$ does not bound any smooth 5-dimensional manifold. This leads naturally to the study of the cobordism groups; we refer to Stong [39] for further details.

Lemma 2.19

1. Let f be a smooth map from a manifold M without boundary to $\mathbb{R}$. Let $L_c := f^{-1}(c)$ be the level sets and let $M_c := f^{-1}(-\infty, c]$. Suppose that f is regular at every point of L_c. Then M_c is a manifold with boundary L_c.

2. Let M be a compact manifold with boundary $\text{bd}(M)$. Take two copies $M_\pm$ of M. Join them at the common boundary to define the *double DM* $:= M_+ \cup_{\text{bd}(M)} M_-$.

   (a) The collar $[0, 1] \times \text{bd}(M)$ is diffeomorphic to a neighborhood of $\text{bd}(M)$ in M.

   (b) The double DM is a manifold.

   (c) $M_\pm$ are smooth manifolds with boundary $\text{bd}(M) = M_+ \cap M_-$.

   (d) There exists a smooth function $f : DM \to \mathbb{R}$ so $\text{bd}(M) = f^{-1}(0)$, so every point of $\text{bd}(M)$ is a regular value, so $M_+ = f^{-1}[0, \infty)$, and so $M_- = f^{-1}(-\infty, 0]$.

3. Let M be a manifold with boundary $\text{bd}(M)$. If $\{(\mathcal{O}_\alpha, \phi_\alpha)\}$ is a coordinate atlas for M which gives M an orientation, then the restriction of the atlas to $\text{bd}(M)$ induces a natural orientation on $\text{bd}(M)$.

Proof. Assertion 1 is an immediate consequence of the proof given of Theorem 2.1. We now prove Assertion 2, which is in essence the converse of Assertion 1. Using a partition of unity, construct a vector field ξ which is non-zero and points inward along $\text{bd}(M)$. Let $\Phi_t(x)$ be the associated flow discussed in Lemma 2.6. The map $\Phi(t, y) := \Phi_t(y)$ provides a diffeomorphism between the collar $[0, \epsilon] \times \text{bd}(M)$ to a neighborhood of $\text{bd}(M)$ in M for some $\epsilon > 0$. Rescaling the parameter t permits us to take $\epsilon = 1$ and establish Assertion 2-a. Gluing $[-1, 0] \times \text{bd}(M)$ in M_- to $[0, 1] \times \text{bd}(M)$ in M_+ along $\{0\} \times \text{bd}(M)$ gives the required smooth structure to DM and establishes Assertion 2-b and Assertion 2-c. We have an identification of $[-1, 1] \times \text{bd}(M)$ with a neighborhood of $\text{bd}(M)$ in DM. We use the collection of functions constructed in the proof of Lemma 1.21 to find a smooth monotonically increasing function ψ taking values in $[-1, 1]$ so that

$$\psi(t) = \begin{cases} \frac{2}{3} & \text{if } t \geq \frac{2}{3} \\ t & \text{if } t \in [-\frac{1}{2}, \frac{1}{2}] \\ -\frac{2}{3} & \text{if } t \leq -\frac{2}{3} \end{cases} .$$

We now define $f(t, y) = \psi(t)$ for $(y, t) \in \mathrm{bd}(M) \times [-1, 1]$. We extend $f(x) = \pm \frac{2}{3}$ on $M_{\pm}$ minus the appropriate collar to construct the desired function f and complete the proof of Assertion 2.

Let M be an oriented manifold. Let $\mathcal{O}_\alpha$ be a coordinate atlas where the coordinate functions satisfy $\det(\Phi'_{\alpha\beta}) > 0$. If $\mathcal{O}_\alpha$ intersects the boundary of M, we may assume $\mathcal{O}_\alpha$ is defined by the relation $x_\alpha^1 \leq 0$. Since the boundary is preserved, $\Phi_{\alpha\beta}^1(0, y) = 0$. Furthermore, $\partial_{x^1} \Phi_{\alpha\beta}^1(0, y) > 0$. We have:

$$
0 \;\leq\; \det \begin{pmatrix} \partial_{x^1} \Phi_{\alpha\beta}^1 & \partial_{x^2} \Phi_{\alpha\beta}^1 & \cdots \\ \partial_{x^1} \Phi_{\alpha\beta}^2 & \partial_{x^2} \Phi_{\alpha\beta}^2 & \cdots \\ & \cdots & \cdots \end{pmatrix} = \det \begin{pmatrix} \partial_{x^1} \Phi_{\alpha\beta}^1 & 0 & \cdots \\ \partial_{x^1} \Phi_{\alpha\beta}^2 & \partial_{x^2} \Phi_{\alpha\beta}^2 & \cdots \\ \cdots & \cdots & \cdots \end{pmatrix}
$$

$$
= \; \partial_{x^1} \Phi_{\alpha\beta}^1 \cdot \det \begin{pmatrix} \partial_{x^2} \Phi_{\alpha\beta}^2 & \partial_{x^3} \Phi_{\alpha\beta}^2 & \cdots \\ \partial_{x^2} \Phi_{\alpha\beta}^3 & \partial_{x^3} \Phi_{\alpha\beta}^3 & \cdots \\ & \cdots & \cdots \end{pmatrix} .
$$

The transition functions for $\mathrm{bd}(M)$ are given by $y \to (\Phi_{\alpha\beta}^2(0, y), \dots, \Phi_{\alpha\beta}^m(0, y))$. Since $\partial_{x^1} \Phi_{\alpha\beta}^1(0, y) > 0$, we conclude from the expression above that

$$
\det \begin{pmatrix} \partial_{x^2} \Phi_{\alpha\beta}^2 & \partial_{x^3} \Phi_{\alpha\beta}^2 & \cdots \\ \partial_{x^2} \Phi_{\alpha\beta}^3 & \partial_{x^3} \Phi_{\alpha\beta}^3 & \cdots \\ \cdots & \cdots & \cdots \end{pmatrix} > 0 . \qquad \square
$$

If Φ is a local diffeomorphism from an open set $\mathcal{U}_1$ of $\mathbb{R}^m$ to another open set $\mathcal{U}_2$ of $\mathbb{R}^m$ and if f is a smooth function with compact support in $\mathcal{U}_2$, then f is integrable on $\mathcal{U}_2$ and the Change of Variable Theorem (see Theorem 1.28) yields

$$
\int_{\mathcal{U}_2} f = \int_{\mathcal{U}_1} (f \circ \Phi) \cdot |\det \Phi'| . \tag{2.3.f}
$$

Let M be compact and oriented; let $\{(\mathcal{O}_\alpha, \phi_\alpha)\}$ be a coordinate atlas giving the orientation for $1 \leq \alpha \leq \ell$. Let Ξ_α be a partition of unity subordinate to this cover. Let ω be a smooth m-form on M. In each coordinate chart, express $\omega = \omega_\alpha (dx_\alpha^1 \wedge \cdots \wedge dx_\alpha^m)$. Then:

$$
\omega_\alpha = \det \left(\frac{\partial x_\alpha^\mu}{\partial x_\beta^\nu} \right) \omega_\beta .
$$

Since M is oriented, the Jacobian determinant is positive and the absolute value of the determinant plays no role. Thus we may use Equation (2.3.f) to show that the following integral is independent of the choices made:

$$
\int_M \omega := \sum_\alpha \int_{\mathcal{U}_\alpha} \Xi_\alpha \omega_\alpha .
$$

2.3.8 GREEN'S THEOREM. The generalized Stokes' Theorem will include the classic theorems of vector calculus as special cases. Let $\gamma : [a, b] \to \mathbb{R}^m$ be a piecewise smooth curve and let $\omega = f_1 dx^1 + \cdots + f_m dx^m$ be a 1-form. The line integral is given by:

$$\int_\gamma \omega := \int_{t=a}^b \left\{ \sum_{i=1}^m f_i(\gamma(t)) \frac{dx^i}{dt} \right\} dt \,.$$

This is independent of the parametrization as long as the orientation is preserved. In this special instance, $df = f'(t)dt$. The *scalar curl* of a 1-form in the plane $\omega = pdx + qdy$ is:

$$\mathrm{sc}(\omega) := \partial_x q - \partial_y p \,.$$

We then have $d\omega = \mathrm{sc}(\omega)dx \wedge dy$. The following result will follow from Theorem 2.22:

Theorem 2.20   (Green's Theorem) *Let R be a compact submanifold of $\mathbb{R}^2$ with smooth boundary. Orient the boundary of R to keep R on the left. Let ω be a smooth 1-form defined on all of R. Then*

$$\int_{\mathrm{bd}(R)} \omega = \int_R \mathrm{sc}(\omega)dxdy \,.$$

2.3.9 THE OPERATORS OF 3-DIMENSIONAL VECTOR CALCULUS. We now introduce the classical operators from vector calculus: div, curl, and grad. Let (x, y, z) be the usual coordinates on an open subset $\mathcal{O}$ of $\mathbb{R}^3$ and, at least formally, let $\nabla := (\partial_x, \partial_y, \partial_z)$. Let $\cdot$ and $\times$ denote the Euclidean inner product and the cross product on $\mathbb{R}^3$. If $f \in C^\infty(\mathcal{O})$, then the *gradient* of f is the smooth vector field given formally by $\mathrm{grad}(f) = \nabla(f)$, i.e.,

$$\mathrm{grad}(f) := (\partial_x f, \partial_y f, \partial_z f) \,.$$

If $F = (f_1, f_2, f_3)$ is a smooth vector field on $\mathcal{O}$, then the *divergence* of F is the smooth function which is given formally by $\mathrm{div}(F) = \nabla \cdot F$; more precisely:

$$\mathrm{div}(F) := \partial_x f_1 + \partial_y f_2 + \partial_z f_3 \,.$$

The *curl* of F is the smooth vector field which is given formally by $\mathrm{curl}(F) = \nabla \times F$, i.e.,

$$\mathrm{curl}(F) := (\partial_y f_3 - \partial_z f_2, \partial_z f_1 - \partial_x f_3, \partial_x f_2 - \partial_y f_1) \,.$$

2.3.10 TRANSLATION TABLE. There is an equivalence of terminology between the language of differential forms and the language of vector calculus in $\mathbb{R}^3$:

1. We identify a 1-form $\omega_1 = f_1 dx + f_2 dy + f_3 dz$ with the vector field $F = (f_1, f_2, f_3)$.

2. We identify a 2-form $\omega_2 = g_1 dy \wedge dz + g_2 dz \wedge dx + g_3 dx \wedge dy$ with the vector field $G = (g_1, g_2, g_3)$.

3. We identify a 3-form $\omega_3 = g\,dx \wedge dy \wedge dz$ with the function g.

4. The exterior derivative $d_0 : C^\infty(\mathcal{O}) \to C^\infty(\Lambda^1(\mathcal{O}))$ corresponds to the gradient:

$$d_0(f) = \partial_x f\,dx + \partial_y f\,dy + \partial_z f\,dz,$$

$$\mathrm{grad}(f) = (\partial_x f, \partial_y f, \partial_z f).$$

5. The exterior derivative $d_1 : C^\infty(\Lambda^1(\mathcal{O})) \to C^\infty(\Lambda^2(\mathcal{O}))$ corresponds to the curl:

$$d_1(f_1 dx + f_2 dy + f_3 dz) = df_1 \wedge dx + df_2 \wedge dy + df_3 \wedge dz$$

$$= (\partial_y f_3 - \partial_z f_2)dy \wedge dz + (\partial_z f_1 - \partial_x f_3)dz \wedge dx + (\partial_x f_2 - \partial_y f_1)dx \wedge dy,$$

$$\mathrm{curl}(f_1, f_2, f_3) = \det \begin{pmatrix} i & j & k \\ \partial_x & \partial_y & \partial_z \\ f_1 & f_2 & f_3 \end{pmatrix}$$

$$= (\partial_y f_3 - \partial_z f_2, \partial_z f_1 - \partial_x f_3, \partial_x f_2 - \partial_y f_1).$$

6. The exterior derivative $d_2 : C^\infty(\Lambda^2(\mathcal{O})) \to C^\infty(\Lambda^3(\mathcal{O}))$ corresponds to the divergence:

$$d(g_1 dy \wedge dz + g_2 dz \wedge dx + g_3 dx \wedge dy) = (\partial_x g_1 + \partial_y g_2 + \partial_z g_3)dx \wedge dy \wedge dz,$$

$$\mathrm{div}(g_1, g_2, g_3) = \partial_x g_1 + \partial_y g_2 + \partial_z g_3.$$

7. The relation $d_1 d_0 = 0$ corresponds to the fact that $\mathrm{curl} \circ \mathrm{grad} = 0$.

8. The relation $d_2 d_1 = 0$ corresponds to the fact that $\mathrm{div} \circ \mathrm{curl} = 0$.

9. If $\Phi : \mathbb{R}^2 \to \mathbb{R}^3$ parametrizes a piece of a smooth surface S, a local unit normal ν can be taken to be given by $\nu := \frac{\partial_u \Phi \times \partial_v \Phi}{\|\partial_u \Phi \times \partial_v \Phi\|}$ and the corresponding element of surface area can be taken to be given by $dA = \|\partial_u \Phi \times \partial_v \Phi\|du\,dv$.

$$\partial_u \Phi \times \partial_v \Phi = \det \begin{pmatrix} i & j & k \\ \partial_u x & \partial_u y & \partial_u z \\ \partial_v x & \partial_v y & \partial_v z \end{pmatrix}$$

$$= (\partial_u y \partial_v z - \partial_u z \partial_v y, \partial_u z \partial_v x - \partial_v z \partial_u x, \partial_u x \partial_v y - \partial_v x \partial_u y),$$

$$F \cdot \nu dA = g_1(\partial_u y \partial_v z - \partial_u z \partial_v y) + g_2(\partial_u z \partial_v x - \partial_v z \partial_u x)$$

$$+ g_3(\partial_u x \partial_v y - \partial_v x \partial_u y)du\,dv,$$

$$dy \wedge dz = (d_u y du + d_v y dv) \wedge (d_u z du + d_v z dv) = (\partial_u y \partial_v z - \partial_u z \partial_v y)du \wedge dv,$$

$$dz \wedge dx = (d_u z du + d_v z dv) \wedge (d_u x du + d_v x dv) = (\partial_u z \partial_v x - \partial_u x \partial_v z)du \wedge dv,$$

$$dx \wedge dy = (d_u x du + d_v x dv) \wedge (d_u y du + d_v y dv) = (\partial_u x \partial_v y - \partial_u y \partial_v x) du \wedge dv,$$

$$g_1 dy \wedge dz + g_2 dz \wedge dx + g_3 dx \wedge dy$$

$$= g_1 (\partial_u y \partial_v z - \partial_u z \partial_v y) du \wedge dv + g_2 (\partial_u z \partial_v x - \partial_v z \partial_u x) du \wedge dv$$

$$+ g_3 (\partial_u x \partial_v y - \partial_v x \partial_u y) du \wedge dv,$$

$$\int_S G \cdot v dA = \int_S \omega_2 .$$

We say S is orientable if we can choose a consistent normal. This defines the "outside" of S; we orient the boundary to keep S on the left standing outside.

2.3.11 THE MÖBIUS STRIP. The Möbius strip is not orientable. We present below two views. The normal line through the center is in gray. It is the dual Möbius strip where we have only drawn the positive part. The Möbius strip itself is striped. The outward unit normal bundle is in gray. Only the positive direction is shown so it does not close.

2.3.12 STOKES' THEOREM AND GAUSS'S THEOREM. The following results will also follow from Theorem 2.22:

Theorem 2.21

1. **(Stokes' Theorem)** *Let S be a smooth compact oriented surface in $\mathbb{R}^3$. Orient the boundary of S to keep S on the left. Let $\omega = f_1 dx + f_2 dy + f_3 dz \in C^\infty(\Lambda^1 S)$. Then*

$$\int_{bd(S)} \omega = \int_S \{\operatorname{curl}(f_1, f_2, f_3) \cdot v\} dS .$$

2. **(Gauss's Theorem)** *Let R be a compact region in $\mathbb{R}^3$ with smooth boundary oriented using the outward normal. Let F be a smooth vector field defined on R. Then*

$$\int_{bd(R)} (F \cdot v) dS = \int_R \operatorname{div} F dx dy dz .$$

The classical Stokes' Theorem seems to be due to Lord Kelvin (William Thompson) and to Sir George Stokes.

Lord Kelvin (1824–1907)   Sir George Stokes (1819–1903)

2.3.13 GENERALIZED STOKES' THEOREM.

Theorem 2.22   (Generalized Stokes' Theorem).   *Let M be a compact smooth oriented manifold of dimension m with smooth boundary* $\mathrm{bd}(M)$. *Let* $\omega_{m-1} \in C^\infty(\Lambda^{m-1}M)$. *Give the boundary* $\mathrm{bd}(M)$ *the orientation discussed in Lemma 2.19. Then*

$$\int_{\mathrm{bd}(M)} \omega_{m-1} = \int_M d\omega_{m-1}.$$

Proof. Let $\mathcal{O}_\alpha$ be a cover of M by coordinate charts. By taking a partition of unity subordinate to this cover, we may restrict to the case that ω_{m-1} is compactly supported in just one coordinate chart $\mathcal{O}$. Thus there exist f_i which are smooth functions compactly supported in $\mathcal{O}$ so that:

$$\omega_{m-1} = \sum_{i=1}^{m} f_i dx^1 \wedge \cdots \wedge dx^{i-1} \wedge dx^{i+1} \wedge \cdots \wedge dx^m,$$

$$d\omega_{m-1} = \sum_{i=1}^{m} (-1)^{i-1} \partial_{x^i} f_i dx^1 \wedge \cdots \wedge dx^m. \tag{2.3.g}$$

We may suppose $\mathcal{O} \subset \mathbb{R}^m_+$. There are two cases that must be considered:

Case I. The coordinate chart $\mathcal{O}$ does not intersect the boundary. This implies

$$\int_{\mathrm{bd}(M)} \omega_{m-1} = 0. \tag{2.3.h}$$

Furthermore, we may let $\int_M d\omega_{m-1}$ range over all of $\mathbb{R}^m$ since the support of ω_{m-1} is contained in $\mathrm{int}(\mathbb{R}^m_+)$. To simplify the notation, we examine the term with $i = m$ in Equation (2.3.g); the other terms are handled similarly modulo an appropriate reordering of the indices. We use Fubini's Theorem (see Theorem 1.19) to express

$$(-1)^{m-1} \int_{\mathcal{O}} (\partial_{x^m} f_m) dx^1 \wedge \cdots \wedge dx^m$$

$$= (-1)^m \int_{-\infty}^{\infty} \left\{ \ldots \left\{ \int_{-\infty}^{\infty} (\partial_{x^m} f_m)(x^1, \ldots, x^m) dx^m \right\} dx^{m-1} \ldots \right\} dx^1.$$

We use the Fundamental Theorem of Calculus together with the fact that f_m has compact support to see:

$$\int_{-\infty}^{\infty} (\partial_{x^m} f_m)(x^1, \dots, x^m) dx^m = f_m(x^1, \dots, x^m)\Big|_{x^m = -\infty}^{x_m = \infty} = 0.$$

This shows

$$(-1)^{m-1} \int_{\mathcal{O}} (\partial_{x^m} f_m) dx^1 \wedge \cdots \wedge dx^m = 0. \tag{2.3.i}$$

We use Equation (2.3.h) and Equation (2.3.i) to see that both sides of Stokes' Theorem vanish in this instance and the result follows in this special case.

Case II. The coordinate chart $\mathcal{O}$ intersects the boundary $\mathbb{R}^{m-1}$. In this case, the integral over $\mathcal{O}$ ranges over $x^1 \in (-\infty, 0]$ and the boundary of M corresponds to $\mathcal{O} \cap \mathbb{R}^{m-1}$; this changes the analysis slightly. If $i > 1$, then $dx^1 \wedge \cdots \wedge dx^{i-1} \wedge dx^{i+1} \wedge \cdots \wedge dx^m$ vanishes on $\mathcal{O} \cap \mathbb{R}^{m-1}$. Consequently, we may use Fubini's Theorem to see:

$$\int_{\mathcal{O} \cap \mathbb{R}^{m-1}} \omega_{m-1} = \int_{-\infty}^{\infty} \left\{ \cdots \left\{ \int_{-\infty}^{\infty} f_1(0, x^2, \dots, x^m) dx^2 \right\} \cdots \right\} dx^m. \tag{2.3.j}$$

The analysis of Case I shows $\int_{\mathcal{O}} (\partial_{x^i} f_i) dx^1 \wedge \cdots \wedge dx^m$ vanishes for $i > 1$. Consequently, after applying Fubini's Theorem and the Fundamental Theorem of Calculus, we have:

$$\begin{aligned}
\int_{\mathcal{O}} d\omega_{m-1} &= \int_{-\infty}^{\infty} \left\{ \cdots \left\{ \int_{-\infty}^{\infty} \left\{ \int_{-\infty}^{0} (\partial_{x^1} f_1(x^1, x^2, \dots, x^m)) dx^1 \right\} dx^2 \right\} \cdots \right\} dx^m \\
&= \int_{-\infty}^{\infty} \left\{ \cdots \left\{ \int_{-\infty}^{\infty} \left\{ f_1(x^1, x^2, \dots, x^m) \Big|_{x^1 = -\infty}^{x^1 = 0} \right\} dx^2 \right\} \cdots \right\} dx^m \\
&= \int_{-\infty}^{\infty} \left\{ \cdots \left\{ \int_{-\infty}^{\infty} f_1(0, x^2, \dots, x^m) dx^2 \right\} \cdots \right\} dx^m.
\end{aligned}$$

This agrees with the integral of Equation (2.3.j) which completes the proof. $\qquad \square$

2.4   APPLICATIONS OF STOKES' THEOREM

Since $d^2 = 0$, we may define the *de Rham cohomology* groups by setting:

$$H_{dR}^k(M) := \frac{\ker\{d : C^\infty(\Lambda^k M) \to C^\infty(\Lambda^{k+1} M)\}}{\mathrm{range}\{d : C^\infty(\Lambda^{k-1} M) \to C^\infty(\Lambda^k M)\}}. \tag{2.4.a}$$

We may use Lemma 2.15 to see that wedge product gives $H_{dR}^*(M) := \oplus_k H_{dR}^k(M)$ the structure of a graded unital skew-commutative ring. Since $dF^* = F^* d$, pullback induces a natural map $F^* : H_{dR}^*(N) \to H_{dR}^*(M)$ that makes de Rham cohomology into a contravariant functor. (We refer to Section 8.1.1 of Book II for further details concerning category theory and functors; the reader need not fuss unduly about this notation.) In Lemma 2.23, we use Stokes' Theorem to exhibit a non-trivial element in the $H_{dR}^m(\mathbb{R}^{m+1} - \{0\})$ that will play a central role in our proof of the Fundamental Theorem of Algebra, of the Brauer Fixed Point Formula, and of the Billiard Ball Theorem subsequently.

Lemma 2.23   Let $\Theta_m \in C^\infty(\Lambda^m(\mathbb{R}^{m+1} - \{0\}))$ be given by

$$\Theta_m(x) := \|x\|^{-m-1} \sum_{i=1}^{m+1} (-1)^{i+1} x^i dx^1 \wedge \cdots \wedge dx^{i-1} \wedge dx^{i+1} \wedge \cdots \wedge dx^{m+1} . \qquad (2.4.b)$$

1. The restriction of Θ to S^m is never vanishing. Thus S^m is orientable.

2. $d\Theta_m = 0$.

3. $\int_{S^m} \Theta_m \neq 0$.

4. $[\Theta_m] \neq 0$ in $H_{dR}^m(\mathbb{R}^{m+1} - \{0\})$ and in $H_{dR}^m(S^m)$.

Proof. Let $\eta_m := \frac{1}{2} d(\| \cdot \|^2) = x^1 dx^1 + \cdots + x^{m+1} dx^{m+1}$. We compute:

$$\begin{aligned}
\eta_m \wedge \Theta_m &= \|x\|^{-m-1} \sum_{i=1}^{m+1} (-1)^{i+1} x_i^2 dx^i \wedge dx^1 \wedge \cdots \wedge dx^{i-1} \wedge dx^{i+1} \wedge \cdots \wedge dx^{m+1} \\
&= \|x\|^{-m+1} dx^1 \wedge \cdots \wedge dx^{m+1} .
\end{aligned}$$

This is non-zero on $\mathbb{R}^{m+1} - \{0\}$ and hence Θ_m is nowhere vanishing on $\mathbb{R}^{m+1} - \{0\}$. Let $\{X_1, \ldots, X_m\}$ be a basis for $T_P S^m$ for some point P in S^m. Choose X_{m+1} so that $\{X_1, \ldots, X_m, X_{m+1}\}$ is a basis for $T_P \mathbb{R}^{m+1}$. Since $\|x\|^2$ is constant on S^m, the restriction of η_m to S^m is zero and hence $\ll \eta_m, X_i \gg = 0$ for $i \leq m$. Extend $\ll \cdot, \cdot \gg$ to a pairing between arbitrary covectors and vectors. We compute:

$$\begin{aligned}
0 \neq & \ll \eta_m \wedge \Theta_m, X_1 \otimes \cdots \otimes X_{m+1} \gg \\
= & \sum_{i=1}^{m+1} (-1)^{i-1} \ll \eta_m, X_i \gg \cdot \ll \Theta_m, X_1 \otimes \ldots \hat{X}_i \cdots \otimes X_{m+1} \gg \\
= & (-1)^m \ll \eta_m, X_{m+1} \gg \ll \Theta_m, X_1 \otimes \cdots \otimes X_m \gg .
\end{aligned}$$

Assertion 1 now follows. To prove Assertion 2, we compute:

$$\begin{aligned}
d\Theta_m &= \left\{ \sum_{i=1}^{m+1} \partial_{x^i} \left(x^i \{(x^1)^2 + \cdots + (x^{m+1})^2\}^{-(m+1)/2} \right) \right\} dx^1 \wedge \cdots \wedge dx^{m+1} \\
&= \left\{ \sum_{i=1}^{m+1} \left(\|x\|^{-(m+1)/2} - (m+1)(x^i)^2 \|x\|^{-(m+3)/2} \right) \right\} dx^1 \wedge \cdots \wedge dx^{m+1} \\
&= \left\{ (m+1)\|x\|^{-(m+1)/2} - (m+1)\|x\|^2 \|x\|^{-(m+3)/2} \right\} dx^1 \wedge \cdots \wedge dx^{m+1} \\
&= 0 .
\end{aligned}$$

Let $\tilde{\Theta}_m = \sum\limits_{i=1}^{m+1} (-1)^{i+1} x^i dx^1 \wedge \cdots \wedge dx^{i-1} \wedge dx^{i+1} \wedge \cdots \wedge dx^{m+1}$. Since $\|x\| = 1$ on S^m,

$$\int_{S^m} \Theta_m = \int_{S^m} \tilde{\Theta}_m \,. \tag{2.4.c}$$

Since $\tilde{\Theta}_m$ is smooth on the unit disk D^{m+1}, we may use Stokes' Theorem to compute

$$\int_{S^m} \tilde{\Theta}_m = \int_{D^{m+1}} d\tilde{\Theta}_m = \int_{D^{m+1}} (m+1) dx^1 \wedge \cdots \wedge dx^{m+1}$$
$$= (m+1)\operatorname{vol}(D^{m+1}) \neq 0 \,. \tag{2.4.d}$$

This proves Assertion 3. Suppose $\Theta_m = d\Psi_{m-1}$ on $\mathbb{R}^{m+1} - \{0\}$ or on S^m. We can apply Stokes' Theorem to compute

$$\int_{S^m} \Theta_m = \int_{\mathrm{bd}(S^m)} \Psi_m = 0 \,. \tag{2.4.e}$$

We combine Equations (2.4.c), (2.4.d), and (2.4.e) to obtain the desired contradiction and complete the proof.   □

2.4.1 HOMOTOPY. We now introduce just a bit of additional notation that will be useful in our discussion of the Fundamental Theorem of Algebra (Theorem 2.25), of the Brauer Fixed Point Formula (Theorem 2.26), and of the Billiard Ball Theorem (Theorem 2.27) subsequently. As these concepts are properly those of algebraic topology, we only present the basic definitions and refer to Spanier [37] for further details.

We say that two smooth maps F_0 and F_1 from a manifold M to a manifold N are *homotopic* if there exists a smooth map Ξ from $M \times [0, 1]$ to N so that $\Xi(x, 0) = F_0(x)$ and so that $\Xi(x, 1) = F_1(x)$. This is an equivalence relation. Suppose N is connected. Fix a base point P_N of N and a base point P_S of the sphere S^k. We say that a map F from S^k to N is *base point preserving* if $F(P_S) = P_N$. A homotopy Ξ between two such maps is said to be base point preserving if $\Xi(P_S, t) = P_N$ for all $t \in [0, 1]$. The kth *homotopy group* $\pi_k(N)$ is the set of base point preserving homotopy classes of maps from S^k to N. This has a natural group structure and up to group isomorphism, $\pi_k(N)$ is independent of the base point. If f is a smooth base point preserving map from M to N, then there is a natural group homomorphism from $\pi_k(M)$ to $\pi_k(N)$.

2.4.2 THE WINDING NUMBER. Let $\Theta_1 \in C^\infty(\mathbb{C} - \{0\})$ be as defined in Equation (2.4.b). If we set $z = x + \sqrt{-1}y$, then

$$\Theta_1 = -\frac{x\,dy - y\,dx}{x^2 + y^2} = \Im\left(\frac{dz}{z}\right) \,.$$

Let γ be a smooth map from $S^1 \to \mathbb{C} - \{0\}$. We define:

$$W(\gamma, 0) := \frac{1}{2\pi} \int_{S^1} \gamma^* \Theta_1 \,.$$

This is the *winding number* of γ about 0, and is a very classical object.

Lemma 2.24

1. If γ_0 and γ_1 are smooth maps from S^1 to $\mathbb{C} - \{0\}$ which are homotopic, then we have that $W(\gamma_0, 0) = W(\gamma_1, 0)$.

2. If $\gamma(z) = z^n$, then $W(\gamma, 0) = n$.

3. The map $\gamma \to W(\gamma, 0)$ is a surjective map from $\pi_1(\mathbb{C} - \{0\})$ to $\mathbb{Z}$.

We note that the winding number is in fact a group homomorphism from $\pi_1(\mathbb{C} - \{0\})$ to $\mathbb{Z}$. This group is Abelian and the base point plays no role. A similar computation using Θ_m would yield a surjective group homomorphism from $\pi_m(\mathbb{R}^{m+1} - \{0\})$ to $\mathbb{Z}$ which is, in fact, an isomorphism. The homotopy groups $\pi_k(S^m)$ vanish for $k < m$. Studying the homotopy groups $\pi_k(\mathbb{R}^{m+1} - \{0\})$ for $k > m$ is a more difficult problem.

Proof. The maps γ_0 and γ_1 are homotopic maps from S^1 to $\mathbb{C} - \{0\}$ implies there is a smooth map $\Gamma : S^1 \times [0, 1] \to \mathbb{C} - \{0\}$ so that $\Gamma(z, 0) = \gamma_0$ and so that $\Gamma(z, 1) = \gamma_1$. We use Stokes' Theorem to prove Assertion 1 by computing:

$$\begin{aligned}
2\pi \{W(\gamma_1, 0) - W(\gamma_0, 0)\} &= \int_{S^1} \gamma_1^* \Theta_1 - \int_{S^1} \gamma_0^* \Theta_1 = \int_{S^1 \times \{1\}} \Gamma^* \Theta_1 - \int_{S^1 \times \{0\}} \Gamma^* \Theta_1 \\
&= \int_{S^1 \times [0,1]} d\Gamma^* \Theta_1 = \int_{S^1 \times [0,1]} \Gamma^* d\Theta_1 = \int_{S^1 \times [0,1]} 0 = 0 \,.
\end{aligned}$$

If $\gamma(z) = z^n$, then $\gamma(\theta) = e^{\sqrt{-1}n\theta} = (\cos(n\theta), \sin(n\theta))$ for $0 \le \theta \le 1$. We prove Assertion 2 by computing:

$$\begin{aligned}
W(\gamma, 0) &= \frac{1}{2\pi} \int_0^{2\pi} \gamma^* \Theta_1 = \frac{1}{2\pi} \int_0^{2\pi} \frac{\cos(n\theta) d\sin(n\theta) - \sin(n\theta) d\cos(n\theta)}{\cos^2(n\theta) + \sin^2(n\theta)} \\
&= \frac{1}{2\pi} \int_0^{2\pi} \{n\cos^2(n\theta)d\theta + n\sin^2(n\theta)d\theta\} = \frac{1}{2\pi} \int_0^{2\pi} n\,d\theta = n \,. \qquad \square
\end{aligned}$$

2.4.3 THE FUNDAMENTAL THEOREM OF ALGEBRA. We can use Stokes' Theorem to establish the following result:

Theorem 2.25   (Fundamental Theorem of Algebra).   *Let $f(z)$ be a complex polynomial of degree $n \geq 1$. Then there exists $z \in \mathbb{C}$ so $f(z) = 0$.*

Proof. Assume to the contrary that $f(z) \neq 0$ for all $z \in \mathbb{C}$. We argue for a contradiction. We may assume without loss of generality that $f(z) = z^n + a_{n-1}z^{n-1} + \cdots + a_0$ is monic. Consider the 1-parameter family of curves with values in $\mathbb{C} - \{0\}$:

$$\gamma_R = \gamma(\theta, R) := f(Re^{\sqrt{-1}\theta}) : S^1 \to \mathbb{C} - \{0\}.$$

By Lemma 2.24,

$$W(\gamma_0, 0) = W(\gamma_R, 0) \tag{2.4.f}$$

is independent of R. Since $\gamma_0(\theta, 0) = a_0$ is the constant path, $\gamma_0^* \Theta_1 = 0$. Consequently

$$W(\gamma_0, 0) = 0. \tag{2.4.g}$$

Let $R := |a_0| + \cdots + |a_{n-1}| + 2$. Define $\tilde{\Gamma} : [0, 2\pi] \times [0, 1] \to \mathbb{C}$ by setting:

$$\begin{aligned}\tilde{\Gamma}(\theta, t) &:= t(Re^{\sqrt{-1}\theta})^n + (1 - t)f(Re^{\sqrt{-1}\theta}) \\ &= R^n e^{\sqrt{-1}n\theta} + (1 - t)a_{n-1}R^{n-1}e^{\sqrt{-1}n\theta} + \cdots + a_0.\end{aligned}$$

We verify that $\tilde{\Gamma}$ takes values in $\mathbb{C} - \{0\}$ by using the triangle inequality to estimate

$$\|\tilde{\Gamma}(\theta, t)\| \geq R^n - |a_{n-1}|R^{n-1} + \cdots + |a_0| \geq R^{n-1}\{R - |a_{n-1}| - \cdots - |a_0|\} > 0.$$

Consequently, γ_R is homotopic to the curve $R^n z^n$. This curve is homotopic to z^n so

$$W(\gamma_R, 0) = n \tag{2.4.h}$$

by Lemma 2.24. We use Equations (2.4.f), (2.4.g), and (2.4.h) to see that $0 = n$ which provides the desired contradiction. $\qquad\square$

2.4.4 BRAUER FIXED POINT THEOREM. This result is due to the Dutch mathematician Luitzen Egbertus Jan Brauer; it follows from Stoke's Theorem.

L. Brauer (1881–1966)

Theorem 2.26 (Brauer Fixed Point Theorem).   *Let $D^m := \{x \in \mathbb{R}^m : \|x\| \le 1\}$ be the unit disk in $\mathbb{R}^m$. If F is a smooth map from D^m to D^m, then there exists a point $P \in D^m$ so $F(P) = P$.*

Proof. We suppose to the contrary that $F(P) \ne P$ for all $P \in D^m$. Let

$$\gamma(t, P) := F(P) + t(P - F(P))$$

be the ray from $F(P)$ to P. Since $\|F(P)\| \le 1$ and $\|P\| \le 1$, $\gamma(t, P)$ belongs to D^m for $t \in [0, 1]$. Let $T(P)$ be the first $t \ge 1$ so that $\|\gamma(t, P)\| = 1$; one may use the quadratic formula to see that the map $P \to T(P)$ is smooth. Let $r(P) := \gamma(T(P), P)$; $r(P)$ is obtained by drawing the half-line from $F(P)$ to P and finding where it intersects the boundary S^{m-1}. This gives a smooth map from D^m to S^{m-1} so that $r(P) = P$ if $P \in S^{m-1}$. We use Lemma 2.23 and Stokes' Theorem to derive a contradiction by computing:

$$0 \ne \int_{S^{m-1}} \Theta_{m-1} = \int_{S^{m-1}} \text{Id}^* \Theta_{m-1} = \int_{S^{m-1}} r^* \Theta_{m-1} = \int_{D^m} dr^* \Theta_{m-1}$$

$$= \int_{D^m} r^* d\Theta_{m-1} = \int_{D^m} 0 = 0. \qquad \square$$

2.4.5 BILLIARD BALL THEOREM. Somewhat fancifully, one thinks of "combing the hair on a billiard ball" as constructing a non-zero tangent vector field on the unit sphere S^m; the vector field gives the direction in which the hair is supposed to lie. This is possible if and only if m is odd.

Theorem 2.27 (Billiard Ball Theorem).   *There exists a smooth nowhere vanishing vector field on S^m if and only if m is odd.*

Proof. Suppose m is odd. We define $F(x^1, \dots, x^{m+1}) = (-x^2, x^1, \dots, -x^{m+1}, x^m)$; this is possible, of course, only if there are an even number of coordinates, i.e., if m is odd. This gives a nowhere vanishing vector field on S^m; this vector field is, of course, nothing but $\sqrt{-1}\vec{z}$ if we identify $\mathbb{R}^{m+1}$ with $\mathbb{C}^{\bar{m}}$ where $2\bar{m} = m + 1$.

We adopt the notation of Lemma 2.23. Let $\Theta_m \in C^\infty(\Lambda^m(\mathbb{R}^{m+1} - \{0\}))$ be given by

$$\Theta_m(x) := \|x\|^{-m-1} \sum_{i=1}^{m+1} (-1)^{i+1} x^i dx^1 \wedge \cdots \wedge dx^{i-1} \wedge dx^{i+1} \wedge \cdots \wedge dx^{m+1}.$$

We showed that $d\Theta_m = 0$ and $\int_{S^m} \Theta_m \ne 0$. Suppose m is even and that there exists a nowhere vanishing vector field. We argue for a contradiction. By replacing F by $\|F\|^{-1}F$, we may assume without loss of generality that F is a unit vector field. Let

$$G(x, \theta) := \cos(\theta) \cdot x + \sin(\theta) \cdot F(x).$$

Since $x \perp F(x)$, this takes values in $S^m \subset \mathbb{R}^{m+1}$ and provides a homotopy as $\theta \in [0, \pi]$ between the identity map and the antipodal map. By Stokes' Theorem:

$$
\int_{S^m} \Theta_m - \int_{S^m} a^* \Theta_m \; = \; \int \mathrm{bd}(S^m \times [0, \pi]) G^* \Theta_m
$$

$$
= \int_{S^m \times [0, \pi]} dG^* \Theta_m = \int_{S^m \times [0, \pi]} G^* d\Theta_m = 0.
$$

As $m + 1$ is odd, $a^* \Theta_m = -\Theta_m$. This implies $2 \int_{S^m} \Theta_m = 0$ which contradicts Lemma 2.23 and establishes the Theorem. □

C H A P T E R 3

Riemannian and
Pseudo-Riemannian Geometry

In Section 3.1, we introduce pseudo-Riemannian geometry. We show that the spheres of even dimension do not admit Lorentzian metrics. We define the pseudo-Riemaⓧare perpendicular nnian measure, verify this agrees with the surface measure dS defined previously in Stokes' Theorem, and compute the volume of spheres and disks. In Section 3.2, we examine connections and their curvature. In Section 3.3, we specialize to the case of the Levi–Civita connection; this is the unique torsion-free Riemannian connection on the tangent bundle. In Section 3.4, we use the Levi–Civita connection to study geodesics. In Section 3.5, we use the Jacobi operator to establish some basic results concerning manifolds of constant sectional curvature. Section 3.6 is devoted to the proof of the Gauss–Bonnet Theorem and the study of Riemann surfaces. In Section 3.7, the generalization to higher dimensions and also to the pseudo-Riemannian setting is exhibited.

3.1   THE PSEUDO-RIEMANNIAN MEASURE

3.1.1 INNER PRODUCT SPACES. We say that the pair $(V, \langle \cdot, \cdot \rangle)$ is an *inner product space* if $\langle \cdot, \cdot \rangle$ is a non-degenerate symmetric bilinear form on a finite-dimensional real vector space V. We say that $0 \neq v \in V$ is *spacelike* (resp. *timelike* or *null*) if $\langle v, v \rangle > 0$ (resp. $\langle v, v \rangle < 0$ or $\langle v, v \rangle = 0$). A subspace W of V is said to be *spacelike* (resp. *timelike* or *null*) if the restriction of $\langle \cdot, \cdot \rangle$ to W is positive definite (resp. negative definite or identically zero).

Lemma 3.1   Let $(V, \langle \cdot, \cdot \rangle)$ be an inner product space.

1. There exists an element of V so $\langle v, v \rangle \neq 0$.

2. There exists an orthogonal direct sum decomposition $V = V_- \oplus V_+$ where V_- is timelike and V_+ is spacelike. Moreover, if $V = \tilde{V}_- \oplus \tilde{V}_+$ is another such decomposition, then we have that $\dim(V_\pm) = \dim(\tilde{V}_\pm)$ and we shall say that $(V, \langle \cdot, \cdot \rangle)$ has signature (p, q) where we set $p = \dim(V_-)$ and $q = \dim(V_+)$.

3. There exist bases $\{e_i^-\}_{1 \leq i \leq p}$ for V_- and $\{e_a^+\}_{1 \leq a \leq q}$ for V_+ so $\langle e_i^-, e_j^- \rangle = -\delta_{ij}$, $\langle e_i^-, e_a^+ \rangle = 0$, and $\langle e_a^+, e_b^+ \rangle = \delta_{ab}$. The set $\{e_1^-, \ldots, e_p^-, e_1^+, \ldots, e_q^+\}$ is said to be an orthonormal basis. Let $v = x_+^a e_a^+ + y_-^i e_i^-$ and $\tilde{v} = \tilde{x}_+^a e_a^+ + \tilde{y}_-^i e_i^-$. Then it follows that $\langle v, \tilde{v} \rangle = x_+^a \tilde{x}_+^a - y_-^i \tilde{y}_-^i$.

4. Any two inner product spaces with the same signature are isomorphic.

Proof. Let $(V, \langle \cdot, \cdot \rangle)$ be an inner product space of dimension m. Suppose Assertion 1 fails, i.e., $\langle v, v \rangle = 0$ for all v. Then

$$0 = \langle v + w, v + w \rangle = \langle v, v \rangle + \langle w, w \rangle + 2 \langle v, w \rangle = 2 \langle v, w \rangle \quad \text{for all} \quad v, w \in V.$$

This contradicts the assumption that $\langle \cdot, \cdot \rangle$ was non-degenerate; Assertion 1 now follows.

If $m = 1$, then $\langle \cdot, \cdot \rangle$ is definite and we are done. We may therefore establish Assertion 2 by induction on m. By Assertion 1, choose $y \in V$ so $\langle y, y \rangle \neq 0$. Let $\Phi(x) := \langle x, y \rangle$ define a non-trivial linear map from V to $\mathbb{R}$; this is non-trivial as $\langle y, y \rangle \neq 0$. Thus $W := \ker\{\Phi\}$ is a linear subspace of V of dimension $m - 1$. Let $0 \neq x \in W$. Suppose $\langle x, w \rangle = 0$ for all $w \in W$. Since $\langle x, y \rangle = 0$ and since $V = y \cdot \mathbb{R} \oplus W$, we conclude $\langle x, \tilde{x} \rangle = 0$ for all $\tilde{x} \in V$. This contradicts the assumption $\langle \cdot, \cdot \rangle$ was non-degenerate. Consequently, the restriction of $\langle \cdot, \cdot \rangle$ to W is non-degenerate. By induction, we may decompose $W = W_+ \oplus W_-$. We obtain a suitable orthogonal decomposition of V by setting:

$$V_+ := \left\{ \begin{array}{ll} y\mathbb{R} \oplus W_+ & \text{if } y \text{ is spacelike} \\ W_+ & \text{if } y \text{ is timelike} \end{array} \right\} \quad \text{and} \quad V_- := \left\{ \begin{array}{ll} y\mathbb{R} \oplus W_- & \text{if } y \text{ is timelike} \\ W_- & \text{if } y \text{ is spacelike} \end{array} \right\}.$$

The decomposition $V = V_- \oplus V_+$ defines projections $\pi_\mp$. Suppose given another decomposition $V = \tilde{V}_- \oplus \tilde{V}_+$. Suppose there exists $0 \neq \xi \in \tilde{V}_-$ so that $\pi_-(\xi) = 0$. Then $\xi \in V_+$ so $\langle \xi, \xi \rangle > 0$ which is false. Thus $\pi_- : \tilde{V}_- \to V_-$ is injective and $\dim(\tilde{V}_-) \leq \dim(V_-)$. Reversing the argument gives $\dim(V_-) \leq \dim(\tilde{V}_-)$. Thus $\dim(V_-) = \dim(\tilde{V}_-)$ and similarly $\dim(V_+) = \dim(\tilde{V}_+)$. This proves Assertion 2. Let $\langle \cdot, \cdot \rangle_\pm$ be the restriction of $\langle \cdot, \cdot \rangle$ to $V_\pm$. We prove Assertion 3 by applying the Gram–Schmidt process to definite inner product spaces $(V_\pm, \langle \cdot, \cdot \rangle_\pm)$; Assertion 4 follows from Assertion 3. $\square$

INDEFINITE FIBER METRICS ON VECTOR BUNDLES. Let V be a vector bundle over M. A *fiber metric* on V of signature (p, q) is a smooth section h to the bundle of symmetric 2-cotensors $S^2(V^*)$ so that the restriction of h to each fiber V_P is a non-degenerate symmetric inner product of signature (p, q) on V_P; the pair (V, h) is then said to be an *orthogonal bundle*. Since the convex combination of positive definite inner products is positive definite, we can use a partition of unity to construct positive definite inner products on V. If $V = TM$, then h is said to be a *pseudo-Riemannian metric* on M. We say h is a *Riemannian metric* if h is positive definite and that h is a *Lorentzian metric* if h has signature $(1, m - 1)$. There always exist Riemannian metrics on M but there are, however, topological restrictions to the construction of indefinite inner products.

Lemma 3.2

1. Let (V, h) be an orthogonal bundle. There exist smooth subbundles $V_\pm$ of V so that h restricts to a spacelike (resp. timelike) fiber metric on V_+ (resp. on V_-), so that V_+ is perpendicular to V_- with respect to h, and so that $V = V_+ \oplus V_-$.

2. If m is even, then TS^m does not admit a fiber metric of indefinite signature. If m is odd, then TS^m admits a Lorentzian metric.

Proof. Let h_e be an auxiliary positive definite fiber metric on V. Since we can diagonalize any quadratic form with respect to a positive definite one, we can find a smooth section Φ to $\mathrm{Hom}(V, V)$ so that $h(v, w) = h_e(\Phi v, w)$. For each $P \in M$, we may diagonalize P; let $\pi_\pm(P)$ be the projections on the positive and negative eigenspaces. The $\pi_\pm$ vary smoothly with P and we use Lemma 2.3 to see that $V_\pm := \mathrm{range}\{\pi_\pm\}$ are smooth subbundles of V; these have the desired properties of Assertion 1.

Let m be even. Suppose to the contrary that TS^m admits a pseudo-Riemannian metric of signature (p, q) for $p > 0$ and $q > 0$. We apply Assertion 1 to decompose $TS^m = V_+ \oplus V_-$ as the orthogonal direct sum of a spacelike and a timelike bundle. Let $H_\pm^m$ be the upper and lower hemispheres of sphere and let P_+ be the north pole and P_- the south pole of S^m. Stereographic projection shows that $S^m - \{P_\pm\}$ is diffeomorphic to $\mathbb{R}^m$ and hence contractible. Consequently, by Lemma 2.3, the bundles $V_\pm$ are trivial over $S^m - \{P_\pm\}$ so we can find non-vanishing smooth sections $\xi_\pm$ to $V_\pm$ over $S^m - \{P_\pm\}$. Use Lemma 1.21 to find smooth functions $\Psi_\pm$ on S^m which are identically 1 on $H_\pm^m$ and which vanish identically near $P_\mp$. Then $\Psi_\pm \xi_\pm$ are smooth tangent vector fields over all of S^m which are non-zero on the hemispheres $H_\pm$. Let $\xi := \Psi_+ \xi_+ \oplus \Psi_- \xi_-$; this is a non-vanishing vector field on S^m. This contradicts Theorem 2.27.

Suppose $m = 2\bar{m} - 1$ is odd. Let $(\cdot, \cdot)$ be the ordinary Euclidean inner product on TS^m. Apply Theorem 2.27 to construct a non-vanishing vector field ξ on S^m. Let $V_- := \xi \cdot \mathbb{R}$ and let $V_+ := V_-^\perp$ be the complementary vector subbundle. Take $(\cdot, \cdot)$ on V_+ and $-(\cdot, \cdot)$ on V_- to construct a Lorentzian metric on S^m. $\qquad\square$

3.1.2 VOLUMES OF SPHERES.

Let $\{(\mathcal{O}_\alpha, (x_\alpha^1, \ldots, x_\alpha^m))\}$ be a coordinate atlas on M. Define the *symmetric tensor product* by setting:

$$dx_\alpha^i \circ dx_\alpha^j := \tfrac{1}{2}(dx_\alpha^i \otimes dx_\alpha^j + dx_\alpha^j \otimes dx_\alpha^i). \tag{3.1.a}$$

Express the metric g on $\mathcal{O}_\alpha$ in the form $g = g_{\alpha,ij} dx_\alpha^i \circ dx_\alpha^j$ where $g_{\alpha,ij} := g(\partial_{x_\alpha^i}, \partial_{x_\alpha^j})$. We define a measure $dv_{g,x}$ on each coordinate chart $\mathcal{O}$ by setting:

$$dv_{g,\alpha} := |\det(g_{\alpha,ij})|^{\frac{1}{2}} dx_\alpha^1 \cdot \cdots \cdot dx_\alpha^m.$$

If g is Riemannian (or equivalently, is positive definite), then $\det(g_{\alpha,ij}) > 0$ and it is not necessary to take the absolute value. Let D_r^m be the ball of radius r in $\mathbb{R}^m$ and let $S_r^{m-1} = \mathrm{bd}(D_r^m)$ be the associated sphere of radius r. Give these manifolds the canonical Riemannian metrics induced from the Euclidean metric on $\mathbb{R}^m$. Define the *Gamma function* for $s > 0$ by setting:

$$\Gamma(s) := \int_0^\infty t^{s-1} e^{-t} dt \, .$$

Lemma 3.3   **(The pseudo-Riemannian measure)** Adopt the notation established above.

1. $dv_{g,\alpha} = dv_{g,\beta}$ on $\mathcal{O}_\alpha \cap \mathcal{O}_\beta$. Thus these local measures patch together to yield an invariantly defined measure dv_g called the pseudo-Riemannian measure on M.

2. Let $\Phi(u,v) := (x(u,v), y(u,v), z(u,v))$ parametrize a surface in $\mathbb{R}^3$. Then $dv_g = dS$ where $dS := \|\partial_u \Phi \times \partial_v \Phi\| du dv$ is as in Stokes' Theorem.

3. $\mathrm{vol}(D_r^m) = \frac{r^m}{m} \mathrm{vol}(S_1^{m-1})$ and $\mathrm{vol}(S_r^{m-1}) = r^{m-1} \mathrm{vol}(S_1^{m-1})$.

4. $\mathrm{vol}(S_1^{m-1}) = 2\pi^{m/2} \Gamma\left(\frac{m}{2}\right)^{-1} = \left\{ \begin{array}{ll} \frac{(2\pi)^{m/2}}{2\cdot4\cdots(m-2)} & \text{if } m \text{ is even} \\ \frac{2(2\pi)^{(m-1)/2}}{1\cdot3\cdots(m-2)} & \text{if } m \text{ is odd} \end{array} \right\}$.

Proof. Let $J_j^i := \frac{\partial x_\alpha^i}{\partial x_\beta^j}$ be the Jacobian matrix. We use the Change of Variable Theorem to see that $dx_\alpha^1 \cdots \cdot dx_\alpha^m = |\det(J)| \, dx_\beta^1 \cdots \cdot dx_\beta^m$. Since $\partial_{x_\beta^j} = J_j^i \partial_{x_\alpha^i}$, $g_{\beta,jk} = J_j^\ell J_k^n g_{\alpha,\ell n}$ and consequently $\det(g_\beta) = \det(J)^2 \det(g_\alpha)$ so $|\det(g_\beta)|^{\frac{1}{2}} = |\det(J)| \, |\det(g_\alpha)|^{\frac{1}{2}}$. Thus we may establish Assertion 1 by computing:

$$\begin{aligned} dv_{g,\alpha} &= |\det(g_\alpha)|^{\frac{1}{2}} dx_\alpha^1 \cdots \cdot dx_\alpha^m = |\det(g_\alpha)|^{\frac{1}{2}} \, |\det(J)| \, dx_\beta^1 \cdots \cdot dx_\beta^m \\ &= |\det(g_\beta)|^{\frac{1}{2}} dx_\beta^1 \cdots \cdot dx_\beta^m = dv_{g,\beta} \, . \end{aligned}$$

If Φ parametrizes a surface in $\mathbb{R}^3$, then:

$$\partial_u \Phi \times \partial_v \Phi = (\partial_u y \cdot \partial_v z - \partial_u z \cdot \partial_v y, \partial_u z \cdot \partial_v x - \partial_u x \cdot \partial_v z, \partial_u x \cdot \partial_v y - \partial_u y \cdot \partial_v x),$$

$$\|\partial_u \Phi \times \partial_v \Phi\|^2 = (\partial_u y \cdot \partial_v z - \partial_u z \cdot \partial_v y)^2 + (\partial_u z \cdot \partial_v x - \partial_u x \cdot \partial_v z)^2$$
$$+ (\partial_u x \cdot \partial_v y - \partial_u y \cdot \partial_v x)^2,$$

$$g_{11}g_{22} - g_{12}g_{12} = \{(\partial_u x)^2 + (\partial_u y)^2 + (\partial_u z)^2\} \cdot \{(\partial_v x)^2 + (\partial_v y)^2 + (\partial_v z)^2\}$$
$$- \{\partial_u x \cdot \partial_v x + \partial_u y \cdot \partial_v y + \partial_u z \cdot \partial_v z\}^2 \, .$$

Assertion 2 follows after a bit of algebraic computation.

Let $F_1(\theta_1, \ldots, \theta_{m-1})$ be a local parametrization of S^{m-1}. Then $F(r, \theta) := rF_1(\theta)$ is a local parametrization of $\mathbb{R}^{m+1}$. We have:

$$\begin{aligned} g(\partial_{\theta_i}, \partial_{\theta_j})(\theta, r) &= r^2 \{(\partial_{\theta_i} F_1) \cdot (\partial_{\theta_j} F_1)\}(\theta) = r^2 g(\partial_{\theta_i}, \partial_{\theta_j})(\theta, 1), \\ g(\partial_{\theta_i}, \partial_r)(\theta, r) &= r\{\partial_{\theta_i} F_1 \cdot F_1\}(\theta) = \tfrac{1}{2} r \partial_{\theta_i} (F_1 \cdot F_1)(\theta) = \tfrac{1}{2} r \partial_{\theta_i}(1) = 0, \\ g(\partial_r, \partial_r)(\theta, r) &= F_1(\theta) \cdot F_1(\theta) = 1 \, . \end{aligned}$$

It now follows that $dv_{\mathbb{R}^m} = r^{m-1} dr dv_{S^{m-1}}$; Assertion 3 now follows. Let

$$\lambda := \int_{-\infty}^{\infty} e^{-x^2} dx \,.$$

In polar coordinates (see Example 1.4.3), we have $dx\,dy = r dr\,d\theta$. Consequently,

$$\lambda^2 = \int_{-\infty}^{\infty} \int_{-\infty}^{\infty} e^{-x^2-y^2} dxdy = \int_0^{\infty} \int_0^{2\pi} re^{-r^2} d\theta dr = -\pi e^{-r^2} \Big|_{r=0}^{\infty} = \pi \,.$$

We use this identity and Assertion 3. We integrate (setting $\tau = r^2$) to see

$$
\begin{aligned}
\pi^{m/2} &= \lambda^m = \int_{\mathbb{R}^m} e^{-\|\vec{x}\|^2} dv_{\mathbb{R}^m} = \int_0^{\infty} r^{m-1} e^{-r^2} dv_{S^{m-1}} dr \\
&= \operatorname{vol}(S^{m-1}) \int_0^{\infty} r^{m-1} e^{-r^2} dr = \tfrac{1}{2} \operatorname{vol}(S^{m-1}) \int_0^{\infty} \tau^{(m-2)/2} e^{-\tau} d\tau \\
&= \tfrac{1}{2} \operatorname{vol}(S^{m-1}) \Gamma\left(\frac{m}{2}\right) \,.
\end{aligned}
$$

This establishes the first identity of Assertion 4; the second then follows from the functional equation $s\Gamma(s) = \Gamma(s+1)$, the fact that $\Gamma(1) = 1$, and the fact that $\Gamma(\tfrac{1}{2}) = \sqrt{\pi}$. □

3.2   CONNECTIONS

We refer to Besse [8] and to Kobayashi and Nomizu [24] for further details concerning the material of this section. Let V be a vector bundle over a manifold M. A connection ∇ on V is a first order partial differential operator from $C^{\infty}(V)$ to $C^{\infty}(T^*M \otimes V)$ satisfying the *Leibnitz formula*:

$$\nabla(fs) = df \otimes s + f\nabla s \quad \text{for} \quad s \in C^{\infty}(V)\,. \tag{3.2.a}$$

Connections always exist locally; if $\vec{s}$ is a local frame for V, then $\nabla(f^i s_i) := df^i \otimes s_i$ defines a connection locally. Since the convex combination of connections is again a connection, we can construct connections using a partition of unity. The associated *directional covariant derivative* $\nabla_X s$ is defined by setting

$$\nabla_X s := \ll X, \nabla s \gg \quad \text{for} \quad s \in C^{\infty}(V) \quad \text{and} \quad X \in C^{\infty}(TM)\,.$$

If $\{e_i\}$ is a basis for TM and if $\{e^i\}$ is the associated *dual basis* for T^*M, the total covariant derivative can be recovered from the directional covariant derivatives by setting

$$\nabla s = e^i \otimes \nabla_{e_i} s\,.$$

We extend ∇ to a *dual connection* ∇^* on V^* by requiring that

$$d \ll s, s^* \gg = \ll \nabla s, s^* \gg + \ll s, \nabla^* s^* \gg \quad \text{for} \quad s \in C^{\infty}(V) \quad \text{and} \quad s^* \in C^{\infty}(V^*)\,.$$

If ∇ and $\tilde{\nabla}$ are connections on vector bundles V and $\tilde{V}$, respectively, then we can define connections on $V \oplus \tilde{V}$ and on $V \otimes \tilde{V}$ by setting, respectively:

$$\nabla(s \oplus \tilde{s}) = \nabla(s) \oplus \tilde{\nabla}(\tilde{s}) \quad \text{and} \quad \nabla(s \otimes \tilde{s}) = \nabla(s) \otimes \tilde{s} + s \otimes \tilde{\nabla}(\tilde{s})\,.$$

3.2.1 HOLONOMY. Let γ be a smooth curve in a manifold M, i.e., a smooth map from some interval $I = [a, b]$ to M. Let V be a vector bundle over M which is equipped with a connection ∇. Let V_0 be the fiber of V over the initial point $\gamma(a)$. A section to V along γ is a smooth map $s : I \to V$ so that $s(t) \in V_{\gamma(t)}$. The section is said to be parallel if $\nabla_{\partial_t} s(t) = 0$. Given a basis $\{e_1, \ldots, e_r\}$ for V_0, we can use the Fundamental Theorem of Ordinary Differential Equations to find a frame $\{e_1(t), \ldots, e_r(t)\}$ for V along γ so $\nabla_{\partial_t} e_i = 0$ and so $e_i(0) = e_i$. If γ is a closed curve, i.e., if $\gamma(a) = \gamma(b)$, then parallel translation along γ defines an invertible linear map T_γ which belongs to $\mathrm{End}(V_0, V_0)$ that is called *holonomy*. The *holonomy group* is the subgroup of $\mathrm{GL}(V_0)$ consisting of all the T_γ as γ ranges over the smooth closed paths in M.

3.2.2 THE CHRISTOFFEL SYMBOLS. Let $\vec{s} = (s_1, \ldots, s_k)$ be a local frame for V and let $\vec{x} = (x^1, \ldots, x^m)$ be a system of local coordinates on M. We may expand

$$\nabla_{\partial_{x^i}} s_a = \Gamma_{ia}{}^b s_b \, .$$

Here the index i ranges from 1 to $m := \dim(M)$ and the indices a, b range from 1 to k, i.e., to the fiber dimension of V. The $\Gamma_{ia}{}^b = {}^\nabla\Gamma_{ia}{}^b$ are referred to as the *Christoffel symbols of the first kind* or sometimes simply as the *Christoffel symbols* of the connection ∇. They are not tensorial. In view of the *Leibnitz formula* given in Equation (3.2.a), ∇ is determined by the Christoffel symbols as we see by computing:

$$\nabla_{\xi^i \partial_{x^i}} (f^a s_a) = \xi^i \{ f^a \Gamma_{ia}{}^b s_b + \partial_{x^i}(f^a) s_a \} \, .$$

Let $\vec{s}\,^* = (s^1, \ldots, s^k)$ be the local *dual frame* field for the dual bundle V^*. The *dual Christoffel symbols* for the *dual connection* on V^* are given by the identity:

$$\nabla^*_{\partial_{x^i}} s^b = -\Gamma_{ia}{}^b s^a \, .$$

Similarly, if V and $\tilde{V}$ are vector bundles which are equipped with connections ∇ and $\tilde{\nabla}$, then the Christoffel symbols of the natural connections on $V \oplus \tilde{V}$ and $V \otimes \tilde{V}$ are given by

$$\Gamma_{ia}{}^b \oplus \tilde{\Gamma}_{i\tilde{a}}{}^{\tilde{b}} \quad \text{and} \quad \Gamma_{ia}{}^b \otimes \mathrm{Id} + \mathrm{Id} \otimes \tilde{\Gamma}_{i\tilde{a}}{}^{\tilde{b}} \, .$$

Let V be equipped with a non-degenerate *fiber metric* h; ∇ is a *Riemannian connection* if

$$dh(s_1, s_2) = h(\nabla s_1, s_2) + h(s_1, \nabla s_2) \quad \text{for} \quad s_i \in C^\infty(V) \, .$$

Equivalently, ∇ is Riemannian if and only if ∇ agrees with ∇^* when we use h to identify V with V^*. We define the *Christoffel symbols of the second kind* by using h to lower indices:

$$\Gamma_{iab} := h(\nabla_{\partial_{x^i}} s_a, s_b) \, .$$

If $\vec{s}$ is a local orthonormal frame for V, then ∇ is Riemannian if and only if

$$\Gamma_{iab} + \Gamma_{iba} = 0 \quad \text{for all} \quad i, a, b \, .$$

3.2.3 THE CURVATURE OPERATOR. The *curvature operator* $\mathcal{R} = {}^{\nabla}\mathcal{R}$ of a connection ∇ is given by:

$$\mathcal{R}(X, Y)s := \{\nabla_X \nabla_Y - \nabla_Y \nabla_X - \nabla_{[X,Y]}\}s \,.$$

Let $\mathcal{R}(\partial_{x^i}, \partial_{x^j})s_a = R_{ija}{}^b s_b$ be the *components of the curvature operator* in a system of local coordinates $\vec{x} = (x^1, \dots, x^m)$ and relative to a local frame $\vec{s} = (s_1, \dots, s_k)$ for V. Then:

$$R_{ija}{}^b = \partial_{x^i}\Gamma_{ja}{}^b - \partial_{x^j}\Gamma_{ia}{}^b + \Gamma_{ic}{}^b \Gamma_{ja}{}^c - \Gamma_{jc}{}^b \Gamma_{ia}{}^c \,. \qquad (3.2.b)$$

We can use a fiber metric to lower indices and define $R(X, Y, s, \tilde{s}) := h(\mathcal{R}(X, Y)s, \tilde{s})$ and $R_{ijab} := h(\mathcal{R}(\partial_{x^i}, \partial_{x^j})s_a, s_b)$.

Lemma 3.4   Let ∇ be a connection on a vector bundle V over M.

1. $\mathcal{R}(X, Y) = -\mathcal{R}(Y, X)$.

2. The curvature is a bundle map from $TM \otimes TM$ to $\mathrm{Hom}(V, V)$ which is given by $\mathcal{R}(\xi^i \partial_{x^i}, \zeta^j \partial_{x^j})f^a s_a = \xi^i \zeta^j f^a R_{ija}{}^b s_b$.

3. If h is a fiber metric on V and if ∇ is Riemannian with respect to h, then $\mathcal{R}$ is skew-symmetric with respect to h, i.e., $h(\mathcal{R}(X, Y)s_1, s_2) + h(s_1, \mathcal{R}(X, Y)s_2) = 0$.

Proof. The symmetry Assertion 1 is immediate. We will show that $\mathcal{R}$ is a $C^\infty(M)$ module homomorphism, i.e., that

$$\mathcal{R}(fX, Y)s = \mathcal{R}(X, fY)s = \mathcal{R}(X, Y)fs = f\mathcal{R}(X, Y)s \,.$$

We may then apply Lemma 2.4 to see that $\mathcal{R}$ is a tensor, i.e., its value only depends on the values of X, Y, and s at the point in question. Assertion 2 will then follow. We compute:

$$
\begin{aligned}
\mathcal{R}(fX, Y) &= \nabla_{fX}\nabla_Y - \nabla_Y\nabla_{fX} - \nabla_{[fX,Y]} \\
&= f\nabla_X\nabla_Y - f\nabla_Y\nabla_X - Y(f)\nabla_X + f\nabla_{[X,Y]} + Y(f)\nabla_X \\
&= f\mathcal{R}(X, Y) \,.
\end{aligned}
$$

Since $\mathcal{R}(X, Y) = -\mathcal{R}(Y, X)$, it follows that $\mathcal{R}(X, fY) = f\mathcal{R}(X, Y)$ as well. We complete the proof of Assertion 2 by verifying:

$$
\begin{aligned}
\mathcal{R}(X, Y)fs &= \{\nabla_X\nabla_Y - \nabla_Y\nabla_X - \nabla_{[X,Y]}\}(fs) \\
&= \nabla_X(f\nabla_Y s) + \nabla_X(Y(f))s - \nabla_Y(f\nabla_X s) \\
&\quad - \nabla_Y(X(f)s) - [X, Y](f)s - f\nabla_{[X,Y]}s \\
&= f\mathcal{R}(X, Y)s + X(f)\nabla_Y s + (XY)(f)s + Y(f)\nabla_X s \\
&\quad - Y(f)\nabla_X s - (YX)(f)s - X(f)\nabla_Y s - [X, Y](f)s \\
&= f\mathcal{R}(X, Y)s \,.
\end{aligned}
$$

Suppose that ∇ is Riemannian. Let $\vec{s}$ be an orthonormal frame field. We must show $R_{ijab} + R_{ijba} = 0$. We use Equation (3.2.b) to compute:

$$
\begin{aligned}
R_{ijab} + R_{ijba} &= \partial_{x^i}\Gamma_{jab} - \partial_{x^j}\Gamma_{iab} + \Gamma_{icb}\Gamma_{ja}{}^c - \Gamma_{jcb}\Gamma_{ia}{}^c \\
&\quad + \partial_{x^i}\Gamma_{jba} - \partial_{x^j}\Gamma_{iba} + \Gamma_{ica}\Gamma_{jb}{}^c - \Gamma_{jca}\Gamma_{ib}{}^c .
\end{aligned}
$$

Since ∇ is Riemannian and the frame field is orthonormal, $\Gamma_{jab} = \Gamma_{jba}$. Consequently, the derivative terms cancel off and we have

$$
R_{ijab} + R_{ijba} = \Gamma_{icb}\Gamma_{ja}{}^c - \Gamma_{jcb}\Gamma_{ia}{}^c + \Gamma_{ica}\Gamma_{jb}{}^c - \Gamma_{jca}\Gamma_{ib}{}^c .
$$

Let $\epsilon_c = h(s_c, s_c) = \pm 1$. We then have $\Gamma_{ia}{}^c = \epsilon_c \Gamma_{iac}$. We use the anti-symmetry in the last two indices to complete the proof:

$$
R_{ijab} + R_{ijba} = \sum_c \epsilon_c \{\Gamma_{icb}\Gamma_{jac} - \Gamma_{jcb}\Gamma_{iac} + \Gamma_{ica}\Gamma_{jbc} - \Gamma_{jca}\Gamma_{ibc}\} = 0 . \qquad \square
$$

3.2.4 THE TORSION TENSOR. If ∇ is a connection on TM, then we may define the *torsion tensor* $\mathcal{T} = {}^\nabla\mathcal{T}$ by:

$$
\mathcal{T}(X,Y) = \nabla_X Y - \nabla_Y X - [X,Y] \quad \text{for} \quad X, Y \in C^\infty(TM) .
$$

One has the following useful fact; it permits one to normalize the choice of the frame so that only the second derivatives of the Christoffel symbols enter into the computation of the curvature:

Lemma 3.5   **(The Torsion Tensor)** Let ∇ be a connection on TM and let $P \in M$.

1. The torsion is a tensor. It defines a bundle map $\mathcal{T} : TM \otimes TM \to TM$ which satisfies the identity $\mathcal{T}(X,Y) = -\mathcal{T}(Y,X)$. It is given by $\mathcal{T}(\xi^i \partial_{x^i}, \eta^j \partial_{x^j}) = \xi^i \eta^j (\Gamma_{ij}{}^k - \Gamma_{ji}{}^k)\partial_{x^k}$.

2. The following conditions are equivalent and if either condition is satisfied at all points of M, then ∇ is said to be an affine connection or a torsion-free connection.

    (a) $\mathcal{T}(X_P, Y_P) = 0$ for all $X_P, Y_P \in T_P M$.

    (b) There exist local coordinates for M centered at P so that $\Gamma_{ij}{}^k(P) = 0$.

3. **(Bianchi Identity)** If ∇ is torsion-free, $\mathcal{R}(X,Y)Z + \mathcal{R}(Y,Z)X + \mathcal{R}(Z,X)Y = 0$.

Proof. It is immediate that $\mathcal{T}(X,Y) = -\mathcal{T}(Y,X)$ and that $\mathcal{T}(\partial_{x^i}, \partial_{x^j}) = (\Gamma_{ij}{}^k - \Gamma_{ji}{}^k)\partial_{x^k}$. We apply Lemma 2.4 to show $\mathcal{T}$ is a tensor and establish Assertion 1 by computing:

$$
\begin{aligned}
\mathcal{T}(\xi X, \eta Y) &= \xi X(\eta)Y + \xi\eta\nabla_X Y - \eta Y(\xi)X - \xi\eta\nabla_Y X \\
&\quad - \xi\eta[X,Y] - \xi X(\eta)Y + \eta Y(\xi)X \\
&= \xi\eta\mathcal{T}(X,Y) .
\end{aligned}
$$

By Assertion 1, $\mathcal{T}(P) = 0$ if and only if $\Gamma_{ij}{}^k(P) = \Gamma_{ji}{}^k(P)$. In particular, if there exists a coordinate system where $\Gamma(P) = 0$, then necessarily $\mathcal{T}$ vanishes at P. Thus Assertion 2-b implies Assertion 2-a. Conversely, assume that Assertion 2-a holds. Choose any system of coordinates $\vec{x} = (x^1, \ldots, x^m)$ on M which are centered at P. Define a new system of coordinates by setting $z^i = x^i + \frac{1}{2} c_{jk}{}^i x^j x^k$ where $c_{jk}{}^i = c_{kj}{}^i$ remains to be chosen. As $\partial_{x^j} = \partial_{z^j} + c_{ji}{}^\ell x^i \partial_{z^\ell}$,

$$\nabla_{\partial_{x^i}} \partial_{x^j}(P) = \nabla_{\partial_{z^i}} \partial_{z^j}(P) + c_{ji}{}^\ell \partial_{z^\ell}(P).$$

Set $c_{ij}{}^\ell := \Gamma_{ij}{}^\ell(P)$; the fact that $c_{ij}{}^\ell = c_{ji}{}^\ell$ is exactly the assumption that the torsion tensor of ∇ vanishes at P. Thus we conclude that the new coordinate system has vanishing Christoffel symbols at P.

Suppose that ∇ is torsion-free. Since $\mathcal{R}$ is a tensor, it suffices to establish the Bianchi identity on a basis. Since ∇ is torsion-free, we may compute:

$$\mathcal{R}(\partial_{x^i}, \partial_{x^j})\partial_{x^k} + \mathcal{R}(\partial_{x^j}, \partial_{x^k})\partial_{x^i} + \mathcal{R}(\partial_{x^k}, \partial_{x^i})\partial_{x^j}$$
$$= \nabla_{\partial_{x^i}} \nabla_{\partial_{x^j}} \partial_{x^k} - \nabla_{\partial_{x^j}} \nabla_{\partial_{x^i}} \partial_{x^k} + \nabla_{\partial_{x^j}} \nabla_{\partial_{x^k}} \partial_{x^i}$$
$$- \nabla_{\partial_{x^k}} \nabla_{\partial_{x^j}} \partial_{x^i} + \nabla_{\partial_{x^k}} \nabla_{\partial_{x^i}} \partial_{x^j} - \nabla_{\partial_{x^i}} \nabla_{\partial_{x^k}} \partial_{x^j}$$
$$= \nabla_{\partial_{x^i}} (\nabla_{\partial_{x^j}} \partial_{x^k} - \nabla_{\partial_{x^k}} \partial_{x^j}) + \nabla_{\partial_{x^j}} (\nabla_{\partial_{x^k}} \partial_{x^i} - \nabla_{\partial_{x^i}} \partial_{x^k})$$
$$+ \nabla_{\partial_{x^k}} (\nabla_{\partial_{x^i}} \partial_{x^j} - \nabla_{\partial_{x^j}} \partial_{x^i}).$$

Since $\nabla_{\partial_{x^j}} \partial_{x^k} - \nabla_{\partial_{x^k}} \partial_{x^j} = [\partial_{x^j}, \partial_{x^k}] = 0$, we have $\nabla_{\partial_{x^i}} (\nabla_{\partial_{x^j}} \partial_{x^k} - \nabla_{\partial_{x^k}} \partial_{x^j}) = 0$; a similar argument shows the remaining terms vanish as well. $\qquad\square$

3.2.5 AFFINE GEOMETRY. An *affine manifold* is a pair (M, ∇) where ∇ is a torsion-free connection on TM. We say that a parametrized curve $\gamma(t)$ in M is a *geodesic* if it satisfies the *geodesic equation*:

$$\nabla_{\dot\gamma} \dot\gamma = 0 \quad \text{or equivalently} \quad \ddot\gamma^i + \Gamma_{jk}{}^i(\gamma(t)) \dot\gamma^j \dot\gamma^k = 0 \qquad (3.2.\text{c})$$

if we express $\gamma = (\gamma^1, \ldots, \gamma^m)$ in some coordinate frame. We say that (M, ∇) is a *complete affine manifold* if every geodesic extends for infinite time.

The following is a useful observation:

Theorem 3.6   *Let (M, ∇) be an affine manifold.*

1. *If $0 \neq c \in \mathbb{R}$ and if $\gamma_c(t) := \gamma(ct)$ is the reparametrized curve, then γ is a geodesic if and only if γ_c is a geodesic.*

2. *Given any point $P \in M$ and any tangent vector $X \in T_P M$, there exists a unique geodesic γ defined on an interval $(-\epsilon, \epsilon)$ for some $\epsilon > 0$ with $\gamma(0) = P$ and $\dot\gamma(0) = X$.*

Proof. Assertion 1 is immediate from the definition; Assertion 2 follows from the Fundamental Theorem of Ordinary Differential Equations. $\qquad\square$

3.3   THE LEVI–CIVITA CONNECTION

Pseudo-Riemannian geometry is to a large extent the study of the Levi–Civita connection. In this section, we will establish the basic facts that we shall need subsequently.

3.3.1 THE FUNDAMENTAL THEOREM OF RIEMANNIAN GEOMETRY. We begin by giving an abstract characterization of the *Levi–Civita connection* [31]. This is named after the Italian mathematician Tullio Levi–Civita.

Tullio Levi–Civita (1873–1941)

Theorem 3.7   *Let (M, g) be a pseudo-Riemannian manifold.*

1. **(Koszul formula)** *If ∇ is a torsion-free Riemannian connection on TM and if $\{X, Y, Z\}$ are vector fields on M, then*

$$2g(\nabla_X Y, Z) = X(g(Y, Z)) + Y(g(X, Z)) - Z(g(X, Y))$$
$$+ g([X, Y], Z) - g([X, Z], Y) - g([Y, Z], X).$$

2. *There exists a unique connection ∇ on the tangent bundle of M (which is called the* Levi–Civita *connection) so that ∇ is torsion-free and Riemannian.*

3. *The curvature of the Levi–Civita connection has the symmetries:*

    (a) $\mathcal{R}(X, Y) + \mathcal{R}(Y, X) = 0.$

    (b) $\mathcal{R}(X, Y)Z + \mathcal{R}(Y, Z)X + \mathcal{R}(Z, X)Y = 0.$

    (c) $g(\mathcal{R}(X, Y)Z, W) + g(Z, \mathcal{R}(X, Y)W) = 0.$

Proof. We prove Assertion 1 by computing:

$$g(\nabla_X Y, Z) = g(\nabla_Y X, Z) + g([X, Y], Z) \qquad \text{(torsion-free)}$$
$$= -g(\nabla_Y Z, X) + Y g(X, Z) + g([X, Y], Z) \qquad \text{(Riemannian)}$$
$$= -g(\nabla_Z Y, X) - g([Y, Z], X) + Y g(X, Z) + g([X, Y], Z) \qquad \text{(torsion-free)}$$
$$= g(\nabla_Z X, Y) - Z g(X, Y) - g([Y, Z], X) + Y g(X, Z) \qquad \text{(Riemannian)}$$
$$+ g([X, Y], Z)$$

$$= g(\nabla_X Z, Y) + g([Z, X], Y) - Zg(X, Y) - g([Y, Z], X) \qquad \text{(torsion-free)}$$
$$+ Yg(X, Z) + g([X, Y], Z)$$

$$= -g(\nabla_X Y, Z) + Xg(Z, Y) + g([Z, X], Y) - Zg(X, Y) \qquad \text{(Riemannian)}$$
$$- g([Y, Z], X) + Yg(X, Z) + g([X, Y], Z).$$

Assertion 1 shows that if the Levi–Civita connection exists, then it is unique. We introduce the notation $g_{ij/k} := \partial_{x^k} g(\partial_{x^i}, \partial_{x^j})$ for the first derivatives of the metric. Applying the Koszul formula to the coordinate frame, then yields the *Christoffel identity* for the *Christoffel symbols*:

$$\Gamma_{ijk} = \tfrac{1}{2}\{g_{jk/i} + g_{ik/j} - g_{ij/k}\}. \qquad (3.3.a)$$

To show that a torsion-free Riemannian connection exists, we use Equation (3.3.a) to define the Christoffel symbols. It is then immediate that $\Gamma_{ijk} = \Gamma_{jik}$ and hence ∇ is torsion-free. Similarly we have that $\Gamma_{ijk} + \Gamma_{ikj} = g_{jk/i}$ and hence ∇ is Riemannian. Since ∇ is unique, this local definition is globally consistent. Assertion 2 follows; Assertion 3 now follows from Lemma 3.4 and from Lemma 3.5. $\qquad \square$

3.3.2 GEOMETRIC REALIZABILITY. Let $(V, \langle \cdot, \cdot \rangle)$ be an inner product space. A 4-tensor $A \in \otimes^4 V^*$ is said to be an *Riemannian algebraic curvature tensor* if A satisfies the symmetries given in Theorem 3.7, i.e.,

$$A(x, y, z, w) + A(y, x, z, w) = 0 \quad \forall\, x, y, z, w \in V,$$
$$A(x, y, z, w) + A(y, z, x, w) + A(z, x, y, w) = 0 \quad \forall\, x, y, z, w \in V,$$
$$A(x, y, z, w) + A(x, y, w, z) = 0 \quad \forall\, x, y, z, w \in V.$$

We say that A is *geometrically realizable* if there exists the germ of a pseudo-Riemannian manifold (M, g), there exists a point P of M and there exists an isometry ϕ from $(T_P M, g_P)$ to $(V, \langle \cdot, \cdot \rangle)$ so that $g(\mathcal{R}_P(x, y)z, w) = A(\phi x, \phi y, \phi z, \phi w)$ for all $x, y, z, w \in T_P M$.

Lemma 3.8   Every Riemannian algebraic curvature tensor on an inner product space $(V, \langle \cdot, \cdot \rangle)$ is geometrically realizable.

Proof. Fix an auxiliary Euclidean metric on V to define $\|x\|$. Let $M_\epsilon := \{x \in V : \|x\| < \epsilon\}$ and let $P = 0$. Let $(x^1, \dots, x^m)$ be the system of local coordinates on V induced by a basis $\{e_i\}$ for V, let $\varepsilon_{ik} := \langle e_i, e_k \rangle$, and let

$$g_{ik} := \varepsilon_{ik} - \tfrac{1}{3} A_{ij\ell k} x^j x^\ell.$$

Clearly $g_{ik} = g_{ki}$. As $g_{ik}(0) = \varepsilon_{ik}$ is non-singular, there exists $\epsilon > 0$ so that g is non-singular on M_ϵ. Let

$$g_{ij/k} = \partial_{x^k} g(\partial_{x^i}, \partial_{x^j}) \quad \text{and} \quad g_{ij/k\ell} := \partial_{x^k} \partial_{x^\ell} g_{ij}.$$

We use Equation (3.2.b) and Equation (3.3.a). Since $g = \varepsilon + O(\|x\|^2)$, $g_{ij/k} = O(\|x\|)$. Consequently we have that $\Gamma = O(\|x\|)$. This implies that:

$$R_{ijk}{}^{\ell} = \partial_{x^i}\Gamma_{jk}{}^{\ell} - \partial_{x^j}\Gamma_{ik}{}^{\ell} + O(\Gamma^2) = \partial_{x^i}\Gamma_{jk}{}^{\ell} - \partial_{x^j}\Gamma_{ik}{}^{\ell} + O(\Gamma^2).$$

Furthermore, $\partial_{x^i}(g_{\ell n}\Gamma_{jk}{}^n) = g_{\ell n}\partial_{x^i}\Gamma_{jk}{}^n + O(\|x\|)$. We lower indices to complete the proof:

$$\begin{aligned}
R_{ijk\ell} &= \partial_{x^i}\Gamma_{jk\ell} - \partial_{x^j}\Gamma_{ik\ell} + O(\|x\|^2), \\
&= \tfrac{1}{2}\{g_{j\ell/ik} + g_{ik/j\ell} - g_{jk/i\ell} - g_{i\ell/jk}\} + O(\|x\|^2) \\
&= \tfrac{1}{6}\{-A_{jik\ell} - A_{jki\ell} - A_{ij\ell k} - A_{i\ell jk} \\
&\quad + A_{ji\ell k} + A_{j\ell ik} + A_{ijk\ell} + A_{ikj\ell}\} + O(\|x\|^2) \\
&= \tfrac{1}{6}\{4A_{ijk\ell} - 2A_{i\ell jk} - 2A_{ik\ell j}\} + O(\|x\|^2) \\
&= A_{ijk\ell} + O(\|x\|^2).
\end{aligned}$$

$\square$

This result shows, as a byproduct, that the symmetries of Theorem 3.7 generate the universal curvature symmetries of the Riemann curvature tensor; there are no additional universal symmetries to be found.

3.3.3 THE SECOND FUNDAMENTAL FORM.

Let M be a submanifold of dimension m of a pseudo-Riemannian manifold (N, g_N) of dimension $m + n$. Let g_M be the restriction of the metric g_N to the submanifold. We suppose that g_M is non-degenerate on M; if this happens, we say M is a *non-degenerate submanifold* of N. This condition is automatic, of course, if g_N is positive definite. Let π_M be orthogonal projection from $T_P N$ to $T_P M$ for $P \in M$. Choose local coordinates $(\vec{x}, \vec{y})$ on an open neighborhood $\mathcal{O}$ of P in N so $\vec{x} = (x^1, \ldots, x^m)$ gives local coordinates on M and so M is defined by $\vec{y} = 0$ where $\vec{y} = (y^1, \ldots, y^n)$. Let ∇^N be the Levi–Civita connection of N. Let $X = a^i(\vec{x})\partial_{x^i} \in C^\infty(TM)$ be a tangent vector field along M. Extend X to a vector field defined on $\mathcal{O}$ by setting $\tilde{V}(\vec{x}, \vec{y}) = a^i(\vec{x})\partial_{x^i}$. If $\tilde{V}_1$ is another extension of X to a vector field on $\mathcal{O}$ so that $\tilde{V}_1 = V$ on M, then the coefficients of $\tilde{V}_1 - \tilde{V}$ relative to the coordinate frame vanish identically on M. Consequently, if $U \in C^\infty(TM)$, then $\nabla^N_U(\tilde{V}_1 - \tilde{V})$ vanishes identically on M. This shows that $\nabla^N_U V := \nabla^N_U \tilde{V}|_M$ is well-defined and independent of the particular extension chosen. Note that

$$\nabla^N_{b^j(x)\partial_{x^j}}(a^i(x)\partial_{x^i}) = b^j(\partial_{x^j}a^i)\partial_{x^i} + a^i b^j \Gamma_{ij}{}^k\partial_{x^k} + a^i b^j \Gamma_{ij}{}^a\partial_{y^a}.$$

Thus if $\Gamma_{ij}{}^a$ is non-trivial, then $\nabla^N_U V$ will in general have components in ∂_{y^a} and thus ∇^N need not induce a connection on TM directly.

Theorem 3.9 *Let (M, g_M) be a non-degenerate submanifold of a pseudo-Riemannian manifold (N, g_N). We may decompose $\nabla^N|_M = \tilde{\nabla}^M + II$ where the projected connection $\tilde{\nabla}^M := \pi_M \nabla^N$ is the Levi–Civita connection of M and where the second fundamental form $II(U, V) := (1 - \pi_M)\nabla^N_U V$ is a symmetric bilinear form taking values in the normal bundle. If U, V, W, and X are tangent to M, then the curvature of N and of M are related by:*

$$R_N(U, V, W, X) = R_M(U, V, W, X) - g(II(V, W), II(U, X)) + g(II(U, W), II(V, X)).$$

Proof. Let X, Y, and Z be tangent vector fields along M. Clearly $\tilde{\nabla}^M_{fY} X = f \tilde{\nabla}^M_Y X$ and clearly $\tilde{\nabla}^M$ is bilinear. We show that $\tilde{\nabla}^M$ is a connection on TM by checking that the Leibnitz rule is satisfied:

$$\pi_M \nabla^N_X(fY) = \pi_M(X(f)Y + f\nabla^N_X Y) = X(f)Y + f\pi_M(\nabla^N_X Y).$$

If $\xi \in T_P N$, then $g_N(Z, \xi) = g_M(Z, \pi_M(\xi))$. Thus $g_N(Z, \nabla_Y X) = g_N(Z, \tilde{\nabla}^M_Y X)$. We show $\tilde{\nabla}^M$ is Riemannian by computing:

$$\begin{aligned}
g_M(\tilde{\nabla}^M_X Y, Z) + g_M(Y, \tilde{\nabla}^M_X Z) &= g_N(\nabla^N_X Y, Z) + g_N(Y, \nabla^N_X Z) \\
&= X g_N(Y, Z) = X g_M(Y, Z).
\end{aligned}$$

We show $\tilde{\nabla}^M$ is torsion-free by using the fact that ∇^N is torsion-free to compute:

$$\tilde{\nabla}^M_X Y - \tilde{\nabla}^M_Y X = \pi_M\left(\nabla_{a^i \partial_{x^i}} b^j \partial_{x^j} - \nabla_{b^j \partial_{x^j}} a^i \partial_{x^i}\right) = \pi_M\{[X, Y]\} = [X, Y].$$

As $\tilde{\nabla}^M$ is a torsion-free and Riemannian connection on TM, Theorem 3.7 implies that $\tilde{\nabla}^M$ is the Levi–Civita connection of M. We show that II is a symmetric tensor by computing:

$$\begin{aligned}
II(U, V) - II(V, U) &= (1 - \pi_M)(\nabla_U V - \nabla_V U) = (1 - \pi_M)([U, V]) = 0, \\
II(fU, V) &= (1 - \pi_M)(\nabla_{fU} V) = f(1 - \pi_M)(\nabla_U V) = f II(U, V), \\
II(U, fV) &= II(fV, U) = f II(V, U) = f II(U, V).
\end{aligned}$$

We complete the proof by computing:

$$\begin{aligned}
R_N(U, V, W, X) &= g(\nabla^N_U \nabla^N_V W, X) - g(\nabla^N_V \nabla^N_U W, X) - g(\nabla^N_{[U,V]} W, X) \\
&= g(\nabla^M_U(\nabla^M_V W + II(V, W)), X) - g(\nabla^M_V(\nabla^M_U W + II(U, W)), X) \\
&\quad - g(\nabla^M_{[U,V]} W, X) \\
&= R_M(U, V, W, X) + U g(II(V, W), X) - g(II(V, W), \nabla^M_U X) \\
&\quad - V g(II(U, W), X) + g(II(U, W), \nabla^N_V X) \\
&= R_M(U, V, W, X) - g(II(V, W), II(U, X)) + g(II(U, W), II(V, X)). \qquad \square
\end{aligned}$$

3.4   GEODESICS

Let (M, g) be a pseudo-Riemannian manifold. A diffeomorphism $T : M \to M$ is said to be an *isometry* if $T^*g = g$. Let ∇ be the Levi–Civita connection of (M, g). Recall that γ is a *geodesic* if $\nabla_{\dot{\gamma}} \dot{\gamma} = 0$.

Lemma 3.10   Let (M, g) be a pseudo-Riemannian manifold.

1. If γ is a geodesic in M, then $|\dot\gamma|$ is constant.

2. If M is a non-degenerate submanifold of an inner product space $(V, \langle \cdot, \cdot \rangle)$ and if g is the restriction of $\langle \cdot, \cdot \rangle$ to M, then a curve γ in M is a geodesic if and only if $\ddot\gamma \perp T_{\gamma(t)} M$ for all t in the parameter space.

3. Let T be an isometry of (M, g). If a unit speed curve is the fixed point set of T, then the curve is a geodesic.

4. Let $(V, \langle \cdot, \cdot \rangle)$ be an inner product space and let $(M_\pm, g) = \{v \in V : \langle v, v \rangle = \pm 1\}$ be a *pseudo-sphere*. Let $P \in M_\pm$ and let Q be a unit vector orthogonal to P. Let

$$\gamma(t) := \left\{ \begin{array}{ll} \cos(t)P + \sin(t)Q & \text{if} \quad \langle P, P \rangle = \langle Q, Q \rangle \\ \cosh(t)P + \sinh(t)Q & \text{if} \quad \langle P, P \rangle = -\langle Q, Q \rangle \end{array} \right\}.$$

Then γ is a geodesic in $M_\pm$ with $\gamma(0) = P$ and $\dot\gamma(0) = Q$.

Proof. Let γ be a geodesic. Since the Levi–Civita connection is Riemannian, we may prove Assertion 1 by computing:

$$\partial_t g(\dot\gamma, \dot\gamma) = 2g(\nabla_{\dot\gamma}\dot\gamma, \dot\gamma) = 0 .$$

If M is a submanifold of N, Theorem 3.9 shows that the Levi–Civita connection of M is obtained by projecting the Levi–Civita connection ∇^N of N onto the tangent space of M. If $N = (V, \langle \cdot, \cdot \rangle)$, then the Christoffel symbols of N vanish. Thus $\nabla^N_{\dot\gamma}\dot\gamma = \ddot\gamma$; Assertion 2 now follows.

Adopt the notation of Assertion 3. Let $P := \sigma(0)$ and $\xi := \dot\sigma(0)$. Let σ_1 be the unique geodesic in M with $\sigma_1(0) = P$ and $\dot\sigma_1(0) = \xi$. Since T fixes σ, $TP = P$ and $T_*\xi = \xi$. Since $T\sigma_1$ is a geodesic starting at P with initial direction ξ, we have $T\sigma_1 = \sigma_1$. Since σ is the fixed point set of T, σ is a reparametrization of σ_1. Since σ has constant non-zero speed, σ is a geodesic.

Adopt the notation of Assertion 4. It is immediate that γ has constant speed and takes values in $M_\pm$. Choose an orthonormal basis $\{e_1, \ldots, e_m\}$ for V so $P = e_1$ and $Q = e_2$. Define an isometry T of $(V, \langle \cdot, \cdot \rangle)$ by setting $T(x^i e_i) := x^1 e_1 + x^2 e_2 - x^3 e_3 - \cdots - x^m e_m$. Then T restricts to an isometry of $M_\pm$ fixing γ pointwise. Assertion 4 now follows from Assertion 3. $\square$

3.4.1 GEODESIC COORDINATES AND THE EXPONENTIAL MAP.

Lemma 3.11   Let P be a point of a Riemannian manifold (M, g). There exists a diffeomorphism $\exp_P$ from a neighborhood $\mathcal{U}$ of 0 in $T_P M$ to a neighborhood $\mathcal{O}$ of P in M so that the curves $t \to \exp_P(t\xi)$ are geodesics in M starting at P with initial direction ξ for $0 \le t \le 1$ and $\xi \in \mathcal{U}$.

Proof. Let $(\mathcal{O}, (x^1, \ldots, x^m))$ be a system of local coordinates on M centered at P. We may then write the geodesic equation in the form given in Equation (3.2.c). The Fundamental Theorem of Ordinary Differential Equations shows that there exists a neighborhood $\mathcal{U}$ of 0 in $T_P M$ and there exists $\epsilon > 0$ so that there is a smooth map

$$F : \mathcal{U} \times [0, \epsilon] \to \mathcal{O}$$

so that the curves $t \to \exp(\xi, t)$ are geodesics starting at P with initial direction ξ. Because the curves $t \to F(\xi, ct)$ are geodesics starting at P with initial direction $c\xi$, $F(\xi, ct) = F(c\xi, t)$. Thus by shrinking $\mathcal{U}$ if need be, we may assume without loss of generality that $\epsilon = 1$ and the curves extend for $0 \le t \le 1$. Let $\exp_P(\xi) := F(\xi, 1)$. This is a smooth map from $\mathcal{U}$ to M with $\exp_P(0) = P$ such that the curves $t \to \exp_P(t\xi)$ are geodesics from P with initial direction ξ. It now follows that $(\exp_P)'(0)\xi = \xi$ so by the Inverse Function Theorem (Theorem 1.8), $\exp_P$ is a diffeomorphism if we shrink $\mathcal{U}$. $\qquad\square$

We now restrict to the Riemannian setting. Fix an orthonormal basis $\{e_1, \ldots, e_m\}$ for $T_P M$. There exists $\epsilon > 0$ so that the map $(x^1, \ldots, x^m) \to \exp_P(x^i e_i)$ is a diffeomorphism from the ball of radius ϵ about the origin in $T_P M$ to a neighborhood of P in M. We use this map to define *geodesic coordinates* on M. Straight lines through the origin are geodesics under this identification.

Lemma 3.12   Let $(x^1, \ldots, x^m)$ be a system of geodesic coordinates on a Riemannian manifold M for $\|x\| < \epsilon$ for some $\epsilon > 0$.

1. Let $\theta(v)$ be a curve in $T_P M$ with $\|\theta\| = 1$. Let $F(r, v) := r\theta(v)$. Then $F_*\partial_r \perp F_*\partial_v$.

2. Let $\xi \in B_\epsilon$. Let $\gamma(t)$ be a curve from 0 to ξ in B_ϵ and let $\tau(t) = t\xi$ for $t \in [0, 1]$. Then $L(\gamma) \ge L(\tau) = \|\xi\|$ with equality if and only if γ is a reparametrization of τ.

3. Let P and Q be two points of a connected Riemannian manifold (M, g). We define the geodesic distance $d_g(P, Q) = \inf_{\gamma:\gamma(0)=P \text{ and } \gamma(1)=Q} L(\gamma)$. This is a metric on M which gives the same underlying topology.

4. Let K be a compact subset of M. There exists $\epsilon = \epsilon(K) > 0$ so that if $P \in K$ and if $d_g(P, Q) < \epsilon(K)$, then there exists a geodesic σ from P to Q of length $d_g(P, Q)$. Furthermore, if $\tilde{\sigma}$ is any curve from P to Q with $d_g(P, Q) = L(\tilde{\sigma})$, then $\tilde{\sigma}$ is a reparametrization of σ.

Proof. Adopt the notation of Assertion 1. As the radial curves from the origin are geodesics, $\nabla_{F_*\partial_r} F_*\partial_r = 0$. Furthermore, $g(F_*\partial_r, F_*\partial_r) = 1$. Finally, $[\partial_r, \partial_v] = 0$. Thus:

$$\partial_r g(F_*\partial_r, F_*\partial_v) = g(\nabla_{F_*\partial_r} F_*\partial_r, F_*\partial_v) + g(F_*\partial_r, \nabla_{F_*\partial_r} F_*\partial_v)$$

$$= 0 + g(F_*\partial_r, \nabla_{F_*\partial_v} F_*\partial_r) = \tfrac{1}{2}\partial_v g(F_*\partial_r, F_*\partial_r) = \tfrac{1}{2}\partial_v(1) = 0.$$

When $r = 0$, $F_*\partial_v = 0$ and thus $g(F_*\partial_r, F_*\partial_v)|_{r=0} = 0$. Thus $g(F_*\partial_r, F_*\partial_v) = 0$, i.e., the radial curves are perpendicular to the angular curves. This proves Assertion 1.

Adopt the notation of Assertion 2. Let $\gamma = r(t)\theta(t)$ where $r(t) \in [0, 1]$ and $\|\theta(t)\| = 1$. Here we may assume $r(t) > 0$ for $t > 0$; if the curve returns to the origin, we simply cut off this part of the curve. We use Assertion 1 to see:

$$\|\dot\gamma\|^2 = g(r'(t)\partial_r, r'(t)\partial_r) + r^2 g(\theta'(t), \theta'(t)) \geq (r'(t))^2 \quad \text{so}$$

$$L(\gamma) = \int \|\dot\gamma\| \geq \int |r'| \geq \int r' = \|\xi\| = L(\tau).$$

If equality holds, there is no angular variance and $r' \geq 0$. Assertion 2 now follows.

Suppose $Q \in \exp_P(B_\epsilon(0))$. If γ is any curve from $P = \exp_P(0)$ to $Q = \exp_P(\xi)$ in M, then either γ stays entirely within $\exp_P(B_\epsilon(0))$, in which case $L(\gamma) \geq |\xi|$, or γ goes outside $\exp_P(B_\epsilon(0))$, in which case the length of the part of γ which goes to the boundary is at least ϵ which is greater than ξ. Thus if $Q \in \exp_P(B_\epsilon(0))$, we conclude $d_g(P, Q) = \|\xi\|$. If Q lies outside $\exp_P(B_\epsilon(0))$, then any curve from P to Q must first reach the boundary of B_ϵ and by Assertion 2 this has length at least ϵ. This shows $d_g(P, Q) \geq 0$ with equality if and only if $P = Q$. It is clear $d_g(P, Q) = d_g(Q, P)$ and $d_g(P, Q) \leq d_g(P, R) + d_g(R, Q)$. Thus d_g is a metric on M. For any P, there exists $\epsilon = \epsilon(P) > 0$ so that if $\delta < \epsilon$ is given, then $\{Q : d_g(P, Q) < \delta\}$ is diffeomorphic to $B_\delta(0)$ in $T_P M$. This shows that the original topology on M and the topology defined by d_g coincide; Assertion 3 now follows. Assertion 4 is immediate from the discussion above.   □

3.4.2 EXAMPLE. By Lemma 3.10, unit speed geodesics in S^m can be parametrized in the form $\cos(t)P + \sin(t)Q$ where P and Q are unit vectors with $P \perp Q$. Fix $P \in S^m$. Let $\{e_1, \ldots, e_{m+1}\}$ be an orthonormal basis for $\mathbb{R}^{m+1}$ where $P = e_1$. Let

$$\xi = x^2 e_2 + \cdots + x^{m+1} e_{m+1} \in T_P S^m.$$

The curve $\gamma(t) := \cos(t)P + \sin(t)\|\xi\|^{-1}\xi$ is a geodesic in S^m from P with initial direction $\|\xi\|^{-1}\xi$. Thus $\tilde\gamma(t) = \cos(\|\xi\|t)P + \sin(\|\xi\|t)\|\xi\|^{-1}\xi$ is a geodesic in S^m from P with initial direction ξ. Setting $t = 1$ yields

$$\exp_P(\xi) = \begin{cases} \cos(\|\xi\|)P + \sin(\|\xi\|)\|\xi\|^{-1}\xi & \text{if } \xi \neq 0 \\ P & \text{if } \xi = 0 \end{cases}.$$

The ball of radius $\frac{\pi}{2}$ about the origin in $T_P M$ is the upper hemisphere in S^m.

Lemma 3.13   If $(x^1, \ldots, x^m)$ are geodesic coordinates on M, then $g_{ij}(0) = \delta_{ij}$, $g_{ij/k}(0) = 0$, and $R_{ijk\ell}(0) = \tfrac{1}{2}\{\partial_{x^i}\partial_{x^k}g_{j\ell} + \partial_{x^j}\partial_{x^\ell}g_{ik} - \partial_{x^i}\partial_{x^\ell}g_{jk} - \partial_{x^j}\partial_{x^k}g_{i\ell}\}$.

Proof. It is immediate from the definition that $g_{ij} = \delta_{ij}$ in geodesic coordinates. We take *coordinate vector fields* $A := a^i \partial_{x^i}$, $B := b^j \partial_{x^j}$, and $C := c^k \partial_{x^k}$ where $\vec{a}$, $\vec{b}$, and $\vec{c}$ belong to $\mathbb{R}^m$. Since straight lines through the origin are geodesics,

$$g(\nabla_A A, C)(0) = 0 \quad \text{for all} \quad A, C. \tag{3.4.a}$$

We polarize Equation (3.4.a). Let $A(t) = A + tB$. If we differentiate with respect to t and set $t = 0$, we see

$$g(\nabla_A B, C)(0) + g(\nabla_B A, C)(0) = 0.$$

Since ∇ is torsion-free, $g(\nabla_A B, C) = g(\nabla_B A, C)$ and hence $g(\nabla_A B, C) = 0$ for any coordinate vector fields. This implies $\Gamma_{ijk}(0) = 0$. We show that the 1-jets of the metric vanish at 0 by computing:

$$g_{jk/i}(0) = \{\Gamma_{ijk} + \Gamma_{ikj}\}(0) = 0.$$

We lower an index in Equation (3.2.b) to compute:

$$
\begin{aligned}
R_{ijk\ell} &= g_{\ell n}(\partial_{x^i}\Gamma_{jk}{}^n + \Gamma_{ip}{}^n\Gamma_{jk}{}^p - \partial_{x^j}\Gamma_{ik}{}^n - \Gamma_{jp}{}^n\Gamma_{ik}{}^p) \\
&= \partial_{x^i}(g_{\ell n}\Gamma_{jk}{}^n) - \partial_{x^j}(g_{\ell n}\Gamma_{ik}{}^n) + O(\|x\|) \\
&= (\partial_{x^i}\Gamma_{jk\ell} - \partial_{x^j}\Gamma_{ik\ell}) + O(\|x\|) \\
&= \tfrac{1}{2}\{\partial_{x^i}\partial_{x^k}g_{j\ell} + \partial_{x^i}\partial_{x^j}g_{k\ell} - \partial_{x^i}\partial_{x^\ell}g_{jk}\} \\
&\quad + \tfrac{1}{2}\{-\partial_{x^j}\partial_{x^k}g_{i\ell} - \partial_{x^j}\partial_{x^i}g_{k\ell} + \partial_{x^j}\partial_{x^\ell}g_{ik}\} + O(\|x\|).
\end{aligned}
$$

We cancel the term $\partial_{x^i}\partial_{x^j}g_{k\ell}$ to complete the proof. $\qquad\square$

3.4.3 GEODESIC CONVEXITY. We say that an open set $\mathcal{O}$ is *geodesically convex* if any two points of $\mathcal{O}$ can be joined by a unique length minimizing geodesic and if this geodesic is contained entirely in $\mathcal{O}$. By Lemma 3.10 and Example 3.4.2, the great circles are the geodesics of the sphere S^m. A hemispherical cap is a geodesic ball; such balls are geodesically convex if and only if the radius is less than $\frac{\pi}{2}$. More generally we have:

Lemma 3.14 Let (M, g) be a Riemannian manifold.

1. Let K be a compact subset of M. There exists $\epsilon = \epsilon(K) > 0$ so that if $P \in K$ and for any $0 < \delta < \epsilon$, then the geodesic ball $B_\delta(P)$ of radius δ about P is geodesically convex.

2. There exists a coordinate atlas for M where the charts $\mathcal{O}_\alpha$ are geodesically convex open sets. In particular, if $\mathcal{O}_{\alpha_1} \cap \cdots \cap \mathcal{O}_{\alpha_\ell}$ is non-empty, then it is contractible. Such a cover is called a *simple cover*.

Proof. Since K is compact, we can choose a uniform $\tilde{\epsilon} > 0$ so that if $P \in K$, then geodesic coordinates $(x_P^1, \ldots, x_P^m)$ which are centered at P are defined for $\|x\| < \tilde{\epsilon}$ on the ball $B_{2\tilde{\epsilon}}(P)$ around P. Furthermore, given any two points Q_1 and Q_2 in $B_{\tilde{\epsilon}}(P)$, then there is a unique shortest geodesic from Q_1 to Q_2 lying in $B_{2\tilde{\epsilon}}(P)$. Again, as K is compact, we may choose C so that $|\Gamma_{ij}{}^k| \leq C$ on $B_{2\tilde{\epsilon}}(P)$. Choose $\epsilon > 0$ so that

$$\epsilon < \tilde{\epsilon} \quad \text{and} \quad \epsilon m^3 C^3 < \tfrac{1}{3}.$$

Suppose Assertion 1 fails. Choose $P \in K$ and points Q_i in $B_\delta(P)$ so that the shortest unit speed geodesic σ from Q_1 and Q_2 does not remain in $B_\delta(P)$. Let $\sigma(t) = (x^1(t), \ldots, x^m(t))$. Then by assumption $d_g(P, \sigma(t))^2 = x^1(t)^2 + \cdots + x^m(t)^2$ has maximum at $t_0 \in (0, T)$. Then:

$$0 \geq \partial_t^2 \left\{ d_g(P, \sigma(t))^2 \right\} \big|_{t=t_0} = 2 \left\{ \dot{x}^i(t) \dot{x}^i(t) + x^i(t) \ddot{x}^i(t) \right\} \big|_{t=t_0}.$$

Because σ is a geodesic, we have the geodesic equation:

$$0 = \nabla_{\dot\sigma} \dot\sigma = \ddot{x}^i(t) + \dot{x}^j(t) \dot{x}^k(t) \Gamma_{jk}{}^i(t).$$

We have $|x^i(t)| \leq 2\delta$ and $|\dot{x}^\ell(t)| \leq 1$. We establish the desired contradiction using the above two equations to see:

$$1 = \dot{x}^i(t) \dot{x}^i(t) \leq 2 |x^i(t) \dot{x}^\ell(t) \Gamma_{jk}{}^\ell(t)| \Big|_{t=t_0} \leq \delta m^3 C \leq \tfrac{2}{3}.$$

This proves Assertion 1; Assertion 2 is an immediate consequence of Assertion 1. $\qquad \square$

3.4.4 THE HOPF–RINOW THEOREM. The following theorem is due to the German mathematicians Heinz Hopf and Willi Ludwig August Rinow.

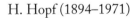
H. Hopf (1894–1971)   W. Rinow (1907–1979)

Theorem 3.15   (Hopf–Rinow [21]).   *Let (M, g) be a connected Riemannian manifold. The following assertions are equivalent and if any is satisfied, then (M, g) is said to be geodesically complete.*

1. *(M, d_g) is a complete metric space.*

2. *All geodesics extend for infinite time.*

3. *There exists a point $P \in M$ so that all geodesics from P extend for infinite time.*

If any of these assertions are satisfied, then any two points P_1 and P_2 of M can be joined by a geodesic σ with $L(\sigma) = d_g(P_1, P_2)$.

Proof. Assume that Assertion 1 holds but, to the contrary, there exists a unit speed geodesic σ defined on $[0, T]$ or $[0, T)$ for $T < \infty$. If it is defined on $[0, T]$, by using the incoming geodesic we can extend it just a bit further to be defined on $[0, T + \epsilon)$ so this is impossible. So we suppose the maximal domain of definition is $[0, T)$ for $T < \infty$. Set $t_i := \frac{i}{i+1}T$ to obtain a sequence $t_i \to t$ and let $P_i = \sigma(t_i)$. Since $d_g(P_i, P_j) \le |t_i - t_j|$, the points $\{P_i\}$ form a Cauchy sequence and converge to a point P of M. Let $B_\epsilon(P)$ be the geodesic ball around P for some $\epsilon > 0$. There exists $\delta > 0$ so that $\sigma(T - \delta, T) \subset B_\epsilon(P)$. We may assume $\delta < \epsilon$. Every geodesic in $B_\epsilon(P)$ extends for time at least ϵ and hence σ extends to time T; this contradiction establishes Assertion 2; it is immediate that Assertion 2 implies Assertion 3.

Assume Assertion 3 holds. Thus all geodesics from P extend for infinite time. Let C be the set of all $t \in \mathbb{R}$ so that if Q is a point of M with $d(P, Q) \le t$, then there exists a geodesic σ from P with $L(\sigma) = d_g(P, Q)$. We use Lemma 3.12 to see C is non-empty. Since $0 < s < t$ and $t \in C$ implies $s \in C$, either $C = [0, \infty)$ or $C = [0, T)$ for some $T < \infty$ or $C = [0, T]$ for some $T < \infty$. We wish to rule out these latter two possibilities.

Suppose $C = [0, T)$ for $T < \infty$. Suppose $d_g(P, Q) = T$ but Q cannot be connected to P by a geodesic of length T. Let γ_n be a sequence of curves from P to Q with $L(\gamma_n) \to T$. Choosing appropriate points on the curves γ_n constructs a sequence of points Q_n with $d_g(P, Q_n) < T$ with $Q_n \to Q$. Express

$$Q_n = \exp_P(\xi_n) \text{ for } \|\xi_n\| = d_g(Q_n, P).$$

We then have $\|\xi_n\| \to d_g(Q, P)$. Consequently, we can find a subsequence so that ξ_{n_k} converges to some element ξ in $T_P M$ with $\|\xi\| = T$. We then have

$$\exp_P(\xi) = \lim_{n \to \infty} \exp_P(\xi_n) = \lim_{n \to \infty} Q_n = Q.$$

Since $\|\xi\| = d_g(P, Q)$, the requisite geodesic from P to Q is given by $\sigma(t) := \exp_P(t\xi)$ for $t \in [0, 1]$; this contradiction shows $C \ne [0, T)$.

Suppose $C = [0, T]$. Then the closed ball $\bar{B}_T(P)$ of radius T about P in M is the image of the closed ball $\bar{B}_T(0)$ of radius T about 0 in $T_P M$ under the exponential map. Since $\bar{B}_T(0)$ is compact, $\bar{B}_T(P)$ is compact. The argument of Lemma 3.12 shows we can find a uniform $\epsilon > 0$ so if Q is any point of $B_T(P)$ and if $d(Q, R) < \epsilon$, then there is a unique shortest geodesic from Q to R of length $d(Q, R)$. Let $d_g(P, Q) < T + \epsilon$. Let γ_i be a sequence of piecewise smooth curves so $L(\gamma_i) \to d_g(P, Q)$. By going sufficiently far out in the sequence, we may assume $L(\gamma_i) < T + \epsilon$ for all i. Let R_i be the first point on γ_i so $d_g(P, R_i) = T$. Decompose γ_i into two curves; one from Q to R_i and one from R_i to P. The curve from R_i to P must have distance at least P and thus the curve from Q to R_i must have distance less than ϵ. By replacing γ_i by the geodesic from Q to R_i and the geodesic from R_i to P, we do not increase the length, and thus we may assume that γ_i is such a broken geodesic. Since the points of distance exactly T from P is a compact set, by passing to a subsequence, we may assume $R_i \to R$. This gives a (possibly) broken geodesic from Q to P of total length $d_g(P, Q)$. If in fact the path is not smooth at R, we can "cut off the leg" and construct a shorter path from Q to P. Since this is not possible, we have constructed a geodesic from Q to P of length $d_g(Q, P)$.

This shows that every point of M can be joined to P by a shortest geodesic of length $d_g(P, Q)$. Let K be a closed bounded subset of M. Since K is bounded, there exists $r > 0$ so that $K \subset \bar{B}_r(P)$. Since $B_r(P) \subset \exp_P(B_r(0))$, $B_r(P)$ is compact. Since K is closed, K is compact. Thus closed bounded subsets of (M, d_g) are compact. It follows (M, d_g) is complete. Thus Assertion 3 implies Assertion 1; the proof that Assertion 3 implies Assertion 1 then shows that if Assertion 2 holds, then any two points can be joined by a shortest geodesic giving the length between the two points. $\qquad\square$

It is worth making a few observations. The shortest geodesic need not be unique; any great circle in the sphere provides a shortest geodesic joining antipodal points. Furthermore, just because there exists a shortest geodesic giving the distance, the space need not be complete; any convex open subset of $\mathbb{R}^n$ has this property. This result can fail in the pseudo-Riemannian setting. The Lie group $\mathrm{SL}(2, \mathbb{R})$ is connected and admits a bi-invariant Lorentzian metric. All geodesics extend for infinite time, but the exponential map is not surjective; we will show in Lemma 6.25 of Book II that there exist points in $\mathrm{SL}(2, \mathbb{R})$ which cannot be connected by a Lorentzian geodesic.

3.5   THE JACOBI OPERATOR

If $X, Y \in T_P M$, let $\mathcal{J}(X) : Y \to \mathcal{R}(Y, X)X$ define the *Jacobi operator*. This operator is named after the German mathematician C. Jacobi.

Carl Gustav Jacob Jacobi (1804–1851)

A vector field X along a geodesic γ is said to be a *Jacobi vector field* if it satisfies the *Jacobi equation*: $\ddot{X} + \mathcal{J}(\dot{\sigma})X = 0$. One says that a smooth map $T : [0, \varepsilon] \times [0, \varepsilon] \to M$ is a *geodesic spray* if T is an embedding such that the curves $\gamma_t(s) := T(t, s)$ are geodesics for all t. The following result (see, for example, do Carmo [9]) provides a geometric motivation for the study of Jacobi vector fields.

Lemma 3.16   Let $T(t, s)$ be a geodesic spray. Then the variation $T_*(\partial_t)$ is a Jacobi vector field along the geodesics $\gamma_t(s) = T(t, s)$.

Proof. Identify ∂_t with $\gamma_* \partial_t$ and ∂_s with $\gamma_* \partial_s$. As the curves $s \to \gamma_t(s)$ are geodesics, we prove the Lemma by computing:

$$0 = \nabla_{\partial_t} \nabla_{\partial_s} \partial_s = \mathcal{R}(\partial_t, \partial_s)\partial_s + \nabla_{\partial_s} \nabla_{\partial_t} \partial_s = \mathcal{J}(\partial_s)\partial_t + \nabla_{\partial_s} \nabla_{\partial_s} \partial_t$$
$$= \mathcal{J}(\dot{\gamma})X + \ddot{X}.$$
$\square$

3.5.1 CONSTANT SECTIONAL CURVATURE.

Let $\{x, y\}$ be a basis for a non-degenerate 2-plane $\pi \subset T_P M$. The *sectional curvature* $\kappa(\pi)$ is defined by setting

$$\kappa(\pi) := \frac{R(x, y, y, x)}{g(x, x)g(y, y) - g(x, y)^2}. \tag{3.5.a}$$

Let $(V, \langle \cdot, \cdot \rangle)$ be an inner product space of signature (p, q). The *pseudo-spheres* are defined by setting

$$S^{\pm} = S^{\pm}(V, \langle \cdot, \cdot \rangle) := \{v \in V : \langle v, v \rangle = \pm 1\}.$$

When considering S^+, we assume that $q > 0$ so there are spacelike vectors; similarly when considering S^-, we assume $p > 0$ to ensure there are timelike vectors.

Lemma 3.17 $S^{\pm}$ is a symmetric space with constant sectional curvature ± 1.

Proof. Fix $P \in S^{\pm}$. Choose an orthonormal basis $\{e_1, \ldots, e_m\}$ for V so that $P = e_1$. Let

$$P^{\perp} := \{\xi \in V : \langle P, \xi \rangle = 0\} = \text{span}\{e_2, \ldots, e_m\}$$

be the perpendicular space. Let $x = (x^2, \ldots, x^m) \in \mathbb{R}^{m-1}$. For x close to 0, define

$$e(x) := x^2 e_2 + \cdots + x^m e_m \in P^{\perp} \quad \text{and} \quad F(x) := (1 \mp \langle x, x \rangle)^{\frac{1}{2}} P + e(x).$$

Then $\langle F(x), F(x) \rangle = \pm(1 \mp \langle x, x \rangle) + \langle x, x \rangle = \pm 1$ so the map $x \to F(x)$ defines a system of local coordinates on $S^{\pm}$ which are centered at P. The Levi–Civita connection of $(V, \langle \cdot, \cdot \rangle)$ is flat. Since P is the normal vector to $S^{\pm}$ at P, $\text{II}(\partial_{x^i}, \partial_{x^j})(P)$ is the projection of $\partial_{x^i} \partial_{x^j} F$ onto P. This shows that

$$\text{II}(\partial_{x^i}, \partial_{x^j})(P) = \mp \langle \partial_{x^i}, \partial_{x^j} \rangle P.$$

Consequently we have (after using Theorem 3.9):

$$R_{S^{\pm}}(U, V, X, W) = \pm\{\langle U, W \rangle \langle V, X \rangle - \langle U, X \rangle \langle V, W \rangle\}.$$

It now follows that $S^{\pm}$ has constant sectional curvature ± 1. Let $Q \in T_P S^{\pm} = P^{\perp}$. By Lemma 3.10, geodesics in $S^{\pm}$ from P take the form

$$\gamma(t) := \left\{ \begin{array}{ll} \cos(t)P + \sin(t)Q & \text{if} \quad \langle P, P \rangle = \langle Q, Q \rangle \\ \cosh(t)P + \sinh(t)Q & \text{if} \quad \langle P, P \rangle = -\langle Q, Q \rangle \end{array} \right\}.$$

Thus the geodesic symmetry which interchanges Q and $-Q$ is induced by the isometry of V which is $+1$ on $P \cdot \mathbb{R}$ and -1 on $P^{\perp}$. This shows the geodesic symmetry is a local isometry and hence $S^{\pm}$ is a local symmetric space.
$\square$

We have the following examples:

1. Let (S^m, g_0) be the unit sphere of $\mathbb{R}^{m+1}$ with the standard positive definite metric. Then

$$R(X, Y, Z, W) = g(X, W)g(Y, Z) - g(X, Z)g(Y, W).$$

This is a Riemannian manifold of constant sectional curvature $+1$. We shall see that any simply connected complete Riemannian manifold of constant sectional curvature $+1$ is isometric to this manifold.

2. Let (H^m, g_0) be the unit pseudo-sphere S^- in a Lorentzian manifold of signature $(1, m)$. Then the induced metric g_0 on H^m is positive definite and

$$R(X, Y, Z, W) = -\{g_0(X, W)g_0(Y, Z) - g_0(X, Z)g_0(Y, W)\}.$$

This has constant sectional curvature -1. We shall see presently that any simply connected complete Riemannian manifold of constant sectional curvature -1 is isometric to this manifold.

3. Let $\vec{x} = (x^1, \ldots, x^m)$ be the usual coordinates on $\mathbb{R}^m$. The following metric on $\mathbb{R}^m$ $(+)$ or on the open unit disk $(-)$ has constant sectional curvature $\pm c$:

$$ds^2 := 4\frac{(dx^1)^2 + \cdots + (dx^m)^2}{(1 \pm \|\vec{x}\|^2)^2}.$$

Let $(V, \langle \cdot, \cdot \rangle)$ be an inner product space. A *Riemannian curvature model* is a triple $(V, \langle \cdot, \cdot \rangle, A)$ where $A \in \otimes^4 V^*$ satisfies the identities of the Riemann curvature tensor given in Theorem 3.7. There are purely algebraic characterizations of constant sectional curvature.

Lemma 3.18   Let $\mathfrak{M} := (V, \langle \cdot, \cdot \rangle, A)$ be a Riemannian curvature model. The following conditions are equivalent and if any is satisfied, then $\mathfrak{M}$ is said to have constant sectional curvature c.

1. $A(x, y, z, w) = c\{\langle x, w \rangle \langle y, z \rangle - \langle x, z \rangle \langle y, w \rangle\}$.

2. $\dfrac{A(e_1, e_2, e_2, e_1)}{\langle e_1, e_1 \rangle \langle e_2, e_2 \rangle - \langle e_1, e_2 \rangle^2} = c$ for any non-degenerate 2-plane $\pi = \text{span}\{e_1, e_2\}$ in V.

3. If $\{e_1, e_2\}$ is an orthonormal set, then $\mathcal{J}(e_1)e_2 = c\langle e_1, e_1 \rangle e_2$.

Proof. It is immediate that Assertion 1 implies Assertion 2. Conversely assume Assertion 2 holds. Let $\{x, y, z\}$ be an orthonormal set of vectors in V. Let $\varepsilon = \langle x, x \rangle \langle y, y \rangle = \pm 1$ and

$$\xi(t) := \left\{ \begin{array}{ll} \cos(t)x + \sin(t)y & \text{if } \varepsilon = 1 \\ \cosh(t)x + \sinh(t)y & \text{if } \varepsilon = -1 \end{array} \right\}.$$

We then have $\langle \xi(t), \xi(t) \rangle = \langle x, x \rangle$. As $\mathfrak{M}$ has constant sectional curvature c,

$$
\begin{aligned}
c\langle x, x \rangle \langle z, z \rangle &= c\langle \xi(t), \xi(t) \rangle \langle z, z \rangle = A(\xi(t), z, z, \xi(t)) \\
&= c\langle x, x \rangle \langle z, z \rangle + 2A(x, z, z, y) \left\{ \begin{array}{ll} \cos(t)\sin(t) & \text{if } \varepsilon = 1 \\ \cosh(t)\sinh(t) & \text{if } \varepsilon = -1 \end{array} \right\}.
\end{aligned}
$$

This shows that $A(x, z, z, y) = 0$. Next suppose that $\{x, y, z, w\}$ is an orthonormal set. Next, polarize the identity $A(x, z, z, y) = 0$ to see $A(x, z, w, y) + A(x, w, z, y) = 0$. Thus:

$$
\begin{aligned}
0 &= A(x, y, z, w) + A(y, z, x, w) + A(z, x, y, w) \\
&= A(x, y, z, w) - A(y, x, z, w) - A(x, z, y, w) \\
&= 3A(x, y, z, w).
\end{aligned}
$$

We have shown that $A(x, z, z, y) = 0$ if $\{x, y, z\}$ is an orthonormal set. We have also shown that $A(x, y, z, w) = 0$ if $\{x, y, z, w\}$ is an orthonormal set. Let $\{e_1, \ldots, e_m\}$ be an orthonormal basis for V. By the curvature symmetries, $A(e_i, e_j, e_k, e_\ell) = 0$ unless $\{i, j\}$ and $\{k, \ell\}$ are distinct. We have $A(e_i, e_j, e_k, e_\ell) = 0$ unless $(i, j) = (k, \ell)$ or $(i, \ell) = (j, k)$. Assertion 1 now follows from Assertion 2.

Let $\{e_1, e_2, e_3\}$ be an orthonormal set of vectors in V. We have shown that $A(e_2, e_1, e_1, e_3) = 0$ so $\mathcal{J}(e_1)e_2 \perp e_3$. Since $A(e_2, e_1, e_1, e_1) = 0$, $\mathcal{J}(e_1)e_2$ is perpendicular to e_1 and e_3. Since e_3 was an arbitrary unit vector perpendicular to e_1 and e_2, we conclude $\mathcal{J}(e_1)e_2$ is some multiple λ of e_2. We show $\lambda = c$ and show Assertion 3 holds by computing:

$$
c\langle e_1, e_1 \rangle \langle e_2, e_2 \rangle = A(e_2, e_1, e_1, e_2) = \langle \mathcal{J}(e_1)e_2, e_2 \rangle = \lambda \langle e_2, e_2 \rangle.
$$

Let $\{e_1, e_2\}$ be an orthonormal set. If Assertion 3 holds, we establish Assertion 1 by checking that $A(e_1, e_2, e_2, e_1) = \langle \mathcal{J}(e_2)e_1, e_1 \rangle = c\langle e_1, e_1 \rangle \langle e_2, e_2 \rangle$. $\qquad \square$

The sectional curvature is a continuous function on the Grassmannian $\mathrm{Gr}_2^0(V)$ of non-degenerate 2-planes in V. In the positive definite setting, this space is compact and, consequently, the sectional curvature is bounded. This fails in the indefinite setting since $\mathrm{Gr}_2^0(V)$ is an open subset of the full Grassmannian $\mathrm{Gr}_2(V)$ and, in fact, the sectional curvature is bounded if and only if it is constant. Furthermore, $R(x, y, y, x) = 0$ if x, y span a degenerate plane if and only if the sectional curvature is constant; see O'Neill [34] for further details.

Let (M, g) be a pseudo-Riemannian manifold. If $(T_P M, g_P, R_P)$ has constant sectional curvature c_P at each point P of M and if $m \geq 3$, then a Schur type lemma shows that c_P is constant. Such manifolds are often also-called *space forms*; if $c > 0$, the manifold is said to be a *spherical space form* while if $c < 0$, it is said to be a *hyperbolic space form*. If $c = 0$, then the manifold is *flat* since the curvature tensor vanishes identically. We refer to Wolf [41] for further information and content ourselves with using Lemma 3.18 to establish the following result:

Lemma 3.19   Let (M, g) and $(\tilde{M}, \tilde{g})$ be two connected pseudo-Riemannian manifolds of constant sectional curvature c and signature (p, q). Let P and $\tilde{P}$ be points of M and $\tilde{M}$. Then there is a local isometry Θ from M to $\tilde{M}$ with $\Theta(P) = \tilde{P}$.

Proof. We use $\exp_P$ to identify a neighborhood of 0 in $T_P M$ with a neighborhood of P in M. Let $g_0 = g_P$. Fix $v \in T_P M$ so that $g_0(v, v) = \varepsilon = \pm 1$. Let $w \in T_P M$ be such that $\{v, w\}$ forms an orthonormal set relative to g_0. We form the geodesic $\gamma(s) := sv$. Let $e(s)$ be a parallel vector field along γ with $e(0) = w$; we regard e as a vector-valued map from a neighborhood of 0 in $\mathbb{R}$ to $T_P M$. Set

$$\theta(v, s) := \left\{ \begin{array}{ll} |c|^{-1} \sin(|c|s)e(s) & \text{if } c\varepsilon > 0 \\ se(s) & \text{if } c = 0 \\ |c|^{-1} \sinh(|c|s)e(s) & \text{if } c\varepsilon < 0 \end{array} \right\} .$$

Let $\{e_1, \ldots, e_m\}$ be an orthonormal basis for $T_P M$ where $e_1 = v$. We wish to show that:

$$g_{sv}(e_i, e_j) = \left\{ \begin{array}{ll} 0 & \text{if } i \neq j \\ g_0(v, v) & \text{if } i = j = 1 \\ s^{-2}\theta^2(v, s)g_0(e_i, e_i) & \text{if } i = j > 1 \end{array} \right\} . \tag{3.5.b}$$

This determines the metric away from the origin and off the *light cone* of null vectors; the metric on the light cone can then be determined by continuity. The Lemma will then follow by choosing an isometry between $T_P M$ and $T_{\tilde{P}} \tilde{M}$; this is possible as M and $\tilde{M}$ have the same signature.

We establish Equation (3.5.b) as follows. Examine the Jacobi vector field $Y := sw$ on γ arising from the geodesic spray $T(t, s) := s(v + tw)$. We use the Levi–Civita connection to covariantly differentiate vector fields. Thus, for example, $\ddot{\gamma} = 0$ since γ is a geodesic and $\ddot{Y} = -\mathcal{J}(\dot{\gamma})Y$ since Y is a Jacobi vector field. We compute:

$$\partial_s \partial_s g_{\gamma(s)}(\dot{\gamma}(s), Y(s)) = \partial_s g_{\gamma(s)}(\ddot{\gamma}(s), Y(s)) + \partial_s g_{\gamma(s)}(\dot{\gamma}(s), \dot{Y}(s))$$
$$= 0 + \partial_s g_{\gamma(s)}(\dot{\gamma}(s), \dot{Y}(s)) = g_{\gamma(s)}(\ddot{\gamma}(s), \dot{Y}(s)) + g(\dot{\gamma}(s), \ddot{Y}(s))$$
$$= 0 - g_{\gamma(s)}(\dot{\gamma}(s), \mathcal{J}(\dot{\gamma})Y(s)) = -R(Y(s), \dot{\gamma}(s), \dot{\gamma}(s), \dot{\gamma}(s)) = 0 .$$

Since $g_{\gamma(s)}(\dot{\gamma}(s), Y(s))(0) = 0$ and $\{\partial_s g_{\gamma(s)}(\dot{\gamma}(s), Y(s))\}(0) = g_0(v, w) = 0$, we may conclude that $Y(s) \perp \dot{\gamma}(s)$ for all s. By Assertion 2 of Lemma 3.18, we have:

$$\ddot{Y} = -\mathcal{J}(\dot{\gamma})Y = -c\varepsilon Y \quad \text{with} \quad Y(0) = 0 \quad \text{and} \quad \dot{Y}(0) = w . \tag{3.5.c}$$

Then $Z(s) := \theta(v, s)e(s)$ satisfies the same ordinary differential equation given in Equation (3.5.c) that is satisfied by Y and, consequently, $Z(s) = Y(s) = sw$. Note that

$$g_{\gamma(s)}(e(s), e(s)) = g(e(0), e(0)) = g_0(w, w) .$$

Thus we have $g_{\gamma(s)}(sw, sw) = \theta^2(v, s)g_0(w, w)$ if $g_0(v, w) = 0$. Equation (3.5.b) now follows by polarization. $\square$

We have actually proved a bit more. By applying the previous lemma to M itself, we have shown that (M, g) is a local two-point homogeneous space, i.e., the local isometries of (M, g) act transitively on the pseudo-sphere bundles

$$S^\pm(M, g) := \{X \in TM : g(X, X) = \pm 1\}.$$

Let $(V, \langle \cdot, \cdot \rangle)$ be an inner product space of signature (p, q). Let $c > 0$. The *pseudo-sphere* $S^\pm(V, \langle \cdot, \cdot \rangle, c) := \{\xi \in V : \langle \xi, \xi \rangle = \pm c^2\}$ discussed previously provide local models for spaces of constant sectional curvature $\pm c$ in all possible signatures.

3.5.2 THE RICCI TENSOR. Let ∇ be the Levi–Civita connection of a Riemannian manifold (M, g). Define the *Ricci tensor* by setting:

$$\rho(X, Y) := \text{Tr}\{Z \to \mathcal{R}(Z, X)Y\}.$$

The curvature symmetries yield $\rho(X, Y) = \rho(Y, X)$. The following result relates the geometry of M to the topology of M through the Ricci tensor:

Theorem 3.20   **(Myers' Theorem [33])** *Let (M, g) be a complete connected Riemannian manifold of dimension $m \geq 2$.*

1. *Let $\sigma(t)$ be a unit speed geodesic in M from P to Q for $0 \leq t \leq b$. Assume that we have the inequality $\rho(\dot\sigma, \dot\sigma) \geq (m-1)(\pi/L)$. If $b > L$, then $d_g(P, Q) < b$.*

2. *Suppose $\rho(X, Y) \geq (m-1)\kappa g(X, Y)$ for some $\kappa > 0$. Then M is compact, the diameter of M is at most $\frac{\pi}{\sqrt{\kappa}}$, and the fundamental group of M is finite.*

Proof. Let $\sigma(t)$ be a unit speed geodesic for $0 \leq t \leq b$ so $d_g(P, Q) = b$. Assume $(\frac{\pi}{b})^2 < L$ i.e., b is larger than predicted by the Lemma so the diameter is larger than expected (or possibly infinite). We argue for a contradiction. Let $Y(t)$ be a perpendicular vector field along σ with $Y(0) = 0$ and $Y(b) = 0$. Consider the variation $T(s, t) = \exp_{\sigma(t)}(sY(t))$. For simplicity of notation, we identify ∂_t with $T_*\partial_t$ and ∂_s with $T_*\partial_s$ since no confusion is likely to ensue. Let

$$L_Y(s) = \int_{t=0}^{b} g(T_*(\partial_t), T_*(\partial_t))^{\frac{1}{2}} dt$$

be the length of the curves $\sigma_s : t \to T(s, t)$. These curves all go from P to Q since $Y(0) = 0$ and $Y(b) = 0$. We compute the first variation of arc length:

$$\partial_s L_Y(s) = \int_{t=0}^{b} \partial_s g(\partial_t, \partial_t)^{\frac{1}{2}} dt = \int_{t=0}^{b} g(\nabla_{\partial_s}\partial_t, \partial_t) g(\partial_t, \partial_t)^{-\frac{1}{2}} dt.$$

Since $Y \perp \dot\sigma$ and since σ is a geodesic when $s = 0$:

$$g(\nabla_{\partial_s}\partial_t, \partial_t)|_{s=0} = g(\nabla_{\partial_t}\partial_s, \partial_t)|_{s=0} = \{\partial_t g(\partial_t, \partial_s) - g(\nabla_{\partial_t}\partial_t, \partial_s)\}|_{s=0} = 0. \quad (3.5.d)$$

Thus, the two relations above show that

$$\partial_s L_Y(s)|_{s=0} = 0.$$

Next, we study the second variation of arc length. We compute:

$$
\begin{aligned}
\partial_s^2 L_Y(s) \quad = \quad & -\int_{t=0}^b g(\nabla_{\partial_s}\partial_t, \partial_t)^2 g(\partial_t, \partial_t)^{-3/2} dt \\
& + \int_{t=0}^b g(\nabla_{\partial_s}\nabla_{\partial_s}\partial_t, \partial_t) g(\partial_t, \partial_t)^{-\frac{1}{2}} dt \\
& + \int_{t=0}^b g(\nabla_{\partial_s}\partial_t, \nabla_{\partial_s}\partial_t) g(\partial_t, \partial_t)^{-\frac{1}{2}} dt.
\end{aligned}
$$

By Equation (3.5.d), $g(\nabla_{\partial_s}\partial_t, \partial_t)|_{s=0} = 0$. Since the curves $s \to T(s, t)$ are geodesics,

$$
\begin{aligned}
\partial_s^2 L_Y(s)|_{s=0} \quad = \quad & \int_{t=0}^b \{g(\nabla_{\partial_s}\nabla_{\partial_t}\partial_s, \partial_t) + g(\nabla_{\partial_t}\partial_s, \nabla_{\partial_t}\partial_s)\} \, dt \\
= \quad & \int_{t=0}^b \{R(\partial_s, \partial_t, \partial_s, \partial_t) - g(\nabla_{\partial_t}\nabla_{\partial_s}\partial_s, \partial_t) + g(\dot{Y}, \dot{Y})\} dt \\
= \quad & \int_{t=0}^b \{g(\dot{Y}, \dot{Y}) - R(Y, \dot{\sigma}, \dot{\sigma}, Y)\} dt.
\end{aligned}
$$

Let $\{e_1, \ldots, e_{m-1}\}$ be parallel perpendicular vector fields along σ. Consider the vector fields $Y_i(t) = \sin(\pi t/b) e_i(t)$. We compute

$$
\begin{aligned}
\sum_{i=1}^{m-1} \partial_s^2 L_{Y_i}(s)|_{s=0} \quad = \quad & \int_{a=0}^b \{(m-1)\cos^2(\pi t/b)(\pi/b)^2 - \sin^2(\pi t/b)\rho(\dot{\sigma}, \dot{\sigma})\} \\
\leq \quad & \int_{a=0}^b \{(m-1)\cos^2(\pi t/b)(\pi/b)^2 - (m-1)L\sin^2(\pi t/b)\} dt \\
< \quad & 0.
\end{aligned}
$$

Thus there exists i so $\partial_s^2 L_{Y_i}(s)|_{s=0} < 0$. As $\partial_s L_{Y_i}(s)|_{s=0} = 0$, $L_{Y_i}(s) < L_{Y_i}(0) = L(\sigma) = b$ for s close to 0. This yields a path of shorter length from P to Q than is provided by σ.

Assume that $\rho(v, v) \geq (m-1)Lg(v, v)$ and (M, g) is complete. We use Assertion 1 to conclude $d(P, Q) \leq L$ for any P and Q in M. Thus $\exp_P : B_L(0) \to M$ is surjective and hence M is compact and we may estimate that $\mathrm{diam}(M) \leq b$. Let $(\tilde{M}, \tilde{g})$ be the universal cover of (M, g). Then $\tilde{\rho}(\tilde{v}, \tilde{v}) \geq (m-1)(\pi/L)\tilde{g}(\tilde{v}, \tilde{v})$ for all $v \in T^*\tilde{M}$. Thus $(\tilde{M}, \tilde{g})$ is compact. It now follows that (M, g) has finite fundamental group. $\qquad\square$

If S^m is the unit sphere of radius r in $\mathbb{R}^{m+1}$, then $\rho(\xi, \xi) = (m-1)r^{-2}\|\xi\|^2$. Since the diameter of S^m is $r\pi$, this result is sharp.

3.6   THE GAUSS–BONNET THEOREM

Let M be a 2-dimensional manifold; if M is orientable, then M is called a *Riemann surface*. Let $g_{ij} := g(\partial_{x^i}, \partial_{x^j})$ represent the metric tensor relative to a system of local coordinates (x^1, x^2) on M. Let $g = \det(g_{ij})^{\frac{1}{2}} = \{g_{11}g_{22} - g_{12}g_{12}\}^{\frac{1}{2}}$ so $|\mathrm{dvol}| = g\,dx^1 dx^2$ is the volume element. The Gaussian curvature K is half the scalar curvature and is the sectional curvature of the only 2-plane. Consequently, $K = g^{-2}R_{1221}$. A system of local coordinates (x^1, x^2) on a Riemann surface is said to be *isothermal* if $ds^2 = e^{2h}((dx^1)^2 + (dx^2)^2)$ for some $h \in C^\infty$. The existence of isothermal coordinates is well-known; we shall postpone the proof as it requires the use of the Newlander–Nirenberg Theorem (we refer to Theorem 4.7 and Section 4.3.1 in Book II). We introduce the notation $h_{/i} := \partial_{x^i} h$ and $h_{/ij} := \partial_{x^i}\partial_{x^j} h$ for the first and second derivatives of h. The expression for the Gaussian curvature is very simple in isothermal coordinates.

Lemma 3.21   If $ds^2 = e^{2h}((dx^1)^2 + (dx^2)^2)$, then $K = -e^{-2h}(h_{/11} + h_{/22})$.

Proof. Let Γ_{ijk} and $\Gamma_{ij}{}^k$ be the Christoffel symbols. We use the Koszul formula of Equation (3.3.a) and then raise indices to see:

$$\Gamma_{111} = h_{/1}e^{2h}, \qquad \Gamma_{112} = -h_{/2}e^{2h}, \quad \Gamma_{121} = \Gamma_{211} = h_{/2}e^{2h},$$
$$\Gamma_{122} = \Gamma_{212} = h_{/1}e^{2h}, \quad \Gamma_{221} = -h_{/1}e^{2h}, \quad \Gamma_{222} = h_{/2}e^{2h},$$
$$\Gamma_{11}{}^1 = h_{/1}, \qquad \Gamma_{11}{}^2 = -h_{/2}, \qquad \Gamma_{12}{}^1 = \Gamma_{21}{}^1 = h_{/2},$$
$$\Gamma_{12}{}^2 = \Gamma_{21}{}^2 = h_{/1}, \qquad \Gamma_{22}{}^1 = -h_{/1}, \qquad \Gamma_{22}{}^2 = h_{/2}.$$

We complete the proof by computing:

$$R_{1221} = g(\nabla_{\partial_{x^1}} \nabla_{\partial_{x^2}} \partial_{x^2}, \partial_{x^1}) - g(\nabla_{\partial_{x^2}} \nabla_{\partial_{x^1}} \partial_{x^2}, \partial_{x^1})$$
$$= g(\nabla_{\partial_{x^1}}\{-h_{/1}\partial_{x^1} + h_{/2}\partial_{x^2}\}, \partial_{x^1}) - g(\nabla_{\partial_{x^2}}\{h_{/2}\partial_{x^1} + h_{/1}\partial_{x^2}\}, \partial_{x^1})$$
$$= e^{2h}\{(-h_{/11} - h_{/1}^2 + h_{/2}^2) - (h_{/22} + h_{/2}^2 - h_{/1}^2)\}. \qquad \square$$

3.6.1 GAUSS THEOREMA EGREGIUM. Let Σ be a smooth 2-dimensional submanifold of $\mathbb{R}^3$ given by the induced metric. Let $T(u^1, u^2) = (x(u^1, u^2), y(u^1, u^2), z(u^1, u^2))$ be an embedding from open subset $\mathcal{O} \subset \mathbb{R}^2$ to $\Sigma \subset \mathbb{R}^3$ that parametrizes part of Σ; i.e., defines a set of local coordinates on Σ. Let '$\times$' be the cross product in $\mathbb{R}^3$. A unit normal to the surface and the second fundamental form II can then be defined by setting

$$\nu := \frac{\partial_{u^1} T \times \partial_{u^2} T}{\|\partial_{u^1} T \times \partial_{u^2} T\|} \quad \text{and} \quad \mathrm{II}(\xi, \eta) = \xi\eta T \cdot \nu$$

where ξ and η are tangent vector fields along $\mathcal{O}$. By Lemma 3.3, $|\mathrm{dvol}| = \|\partial_u T \times \partial_v T\|\,du\,dv$. By Theorem 3.9, $\nabla_\xi \eta = (\xi\eta)T - \mathrm{II}(\xi, \eta) \cdot \nu$ where ∇ is the Levi–Civita connection. Theorem 3.9 then permits to relate the external and the internal geometry. The following result is now immediate:

Theorem 3.22   (Gauss Theorema Egregium).   *If Σ is a surface in $\mathbb{R}^3$, then $K = \frac{\det(II)}{\det(g)}$.*

3.6.2 THE PLANE. Let $(M, g) = (\mathbb{R}^2, g_e)$ be the plane with the usual flat Euclidean metric $g_e := (dx^1)^2 + (dx^2)^2$. Since $g_{ij} = \delta_{ij}$, the Christoffel symbols vanish and $K = 0$. If we parametrize the plane as $T(u, v) = (u, v, 0)$, then we can also show $K = 0$ by computing:

$$\partial_u T = (1, 0, 0), \quad \partial_v T = (0, 1, 0), \quad \nu = (0, 0, 1),$$
$$g_{uu} = 1, \quad g_{uv} = 0, \quad g_{vv} = 1,$$
$$II_{uu} = 0, \quad II_{uv} = 0, \quad II_{vv} = 0.$$

3.6.3 THE CYLINDER. Let $T(u, v) = (\cos u, \sin u, v)$ for $0 \le u \le 2\pi$ and $0 \le v \le 1$ parametrize the cylinder. We show $K = \det(II)/\det(g) = 0$ by computing:

$$\partial_u T = (-\sin u, \cos u, 0), \quad \partial_v T = (0, 0, 1), \quad \nu = (\cos(u), \sin(u), 0),$$
$$g_{uu} = 1, \quad g_{uv} = 0, \quad g_{vv} = 1,$$
$$II_{uu} = -1, \quad II_{uv} = 0, \quad II_{vv} = 0.$$

The cylinder

3.6.4 THE TORUS OF REVOLUTION. We parametrize the unit circle in the xy plane by setting $\gamma(u) = (\cos(u), \sin(u), 0)$. Let $r < 1$ and let S be the surface obtained by taking the set of points at distance r from $\gamma(u)$ in the plane perpendicular to $\dot\gamma$ centered at γ. Alternatively, we can take the circle $x = (1 + r\cos(v))$ and $z = r\sin(v)$ in the xz plane and revolve it around the z-axis. We may parametrize S in the form:

$$\begin{aligned} T(u, v) &= (1 + r\cos(v))\gamma(u) + r\sin(v)e_3 \\ &= ((1 + r\cos(v))\cos(u), (1 + r\cos(v))\sin(u), r\sin(v)) \, ; \end{aligned}$$

$\{\gamma, \dot\gamma, e_3\}$ is a moving orthonormal frame with $\gamma \times \dot\gamma = e_3$, $\dot\gamma \times e_3 = \gamma$, $e_3 \times \gamma = \dot\gamma$. We show that $K = \cos(v)(1 + r\cos(v))^{-1}$ by computing:

$$\partial_u T = (1 + r\cos(v))\dot\gamma, \qquad \partial_v T = -r\sin(v)\gamma + r\cos(v)e_3, \quad \nu = \cos(v)\gamma + \sin(v)e_3,$$
$$g_{uu} = (1 + r\cos(v))^2, \qquad g_{vv} = r^2, \qquad g_{uv} = 0,$$
$$\partial_u^2 T = -(1 + r\cos(v))\gamma, \qquad \partial_v^2 T = -r\cos(v)\dot\gamma - r\sin(v)e_3, \quad \partial_u\partial_v T = -r\sin(b)\dot\gamma$$
$$II_{uu} = -\cos(v)(1 + r\cos(v)), \quad II_{vv} = -r, \qquad II_{uv} = 0.$$

The torus

3.6.5 THE UNIT SPHERE IN $\mathbb{R}^3$.

Let S^2 be the unit sphere in $\mathbb{R}^3$ with the induced metric. The orthogonal group $O(3)$ acts transitively on S^2 by isometries; thus S^2 has constant Gauss curvature. Let $T(x, y) = (x, y, (1 - x^2 - y^2)^{\frac{1}{2}})$ parametrize the upper hemisphere as a graph over the unit disk. We compute:

$$\partial_x T = (1, 0, -x(1 - x^2 - y^2)^{-\frac{1}{2}}), \quad \partial_y T = (0, 1, -y(1 - x^2 - y^2)^{-\frac{1}{2}}),$$
$$g_{xx} = 1 + x^2(1 - x^2 - y^2)^{-1}, \qquad g_{xy} = xy(1 - x^2 - y^2)^{-1},$$
$$g_{yy} = 1 + y^2(1 - x^2 - y^2)^{-1}.$$

We evaluate at $(0, 0)$ to show that $K(0, 0, 1) = 1$ and hence $K \equiv 1$ by computing:

$$g_{11}(0,0) = 1, \qquad\qquad g_{12}(0,0) = 0, \qquad\qquad g_{22}(0,0) = 1,$$
$$\partial_x T(0,0) = (1,0,0), \quad \partial_y T(0,0) = (0,1,0), \quad \nu(0,0) = (0,0,1),$$
$$\partial_x^2 T(0,0) = (0,0,-1), \quad \partial_x\partial_y(0,0) = (0,0,0), \quad \partial_y^2 T(0,0) = (0,0,-1),$$
$$\mathrm{II}_{11}(0,0) = -1, \qquad \mathrm{II}_{12}(0,0) = 0, \qquad \mathrm{II}_{22}(0,0) = -1.$$

The sphere

Stereographic projection permits us to identify the sphere minus the north pole with the plane. Let $T_1(x, y) = t(x, y, 0) + (1 - t)(0, 0, 1)$ parametrize the line between the north pole and a point of the plane. We set $1 = \|T_1\|^2 = t^2(x^2 + y^2) + (1 - t)^2$ to see $t = \frac{2}{x^2+y^2+1}$ and thus $T_1(x, y) = (2x, 2y, x^2 + y^2 - 1)/(1 + x^2 + y^2)$.

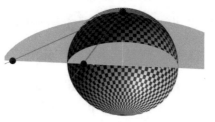

Stereographic projection

Let $\mathcal{E} := (1 + x^2 + y^2)^2$. We compute:

$$\partial_x T_1 = (2(1 + x^2 + y^2) - 4x^2, -4xy, 2x(1 + x^2 + y^2) - 2x(x^2 + y^2 - 1))/\mathcal{E}$$
$$= (2 + 2y^2 - 2x^2, -4xy, 4x)/\mathcal{E},$$
$$\partial_y T_1 = (-4xy, 2(1 + x^2 + y^2) - 4y^2, 2y(1 + x^2 + y^2) - 2y(x^2 + y^2 - 1))/\mathcal{E}$$
$$= (-4xy, 2 + 2x^2 - 2y^2, 4y)/\mathcal{E},$$
$$g_{11} = (4 + 4y^4 + 4x^4 + 8y^2 - 8x^2 - 8x^2y^2 + 16x^2y^2 + 16x^2)/(1 + x^2 + y^2)^4$$
$$= 4(1 + x^2 + y^2)^2/(1 + x^2 + y^2)^4, \qquad g_{12} = 0,$$
$$g_{22} = (4 + 4y^4 + 4x^4 + 8x^2 - 8y^2 - 8x^2y^2 + 16x^2y^2 + 16y^2)/(1 + x^2 + y^2)^4$$
$$= 4(1 + x^2 + y^2)^2/(1 + x^2 + y^2)^4.$$

This yields isothermal coordinates so that

$$ds^2 = 4(dx^2 + dy^2)(1 + x^2 + y^2)^{-2} \tag{3.6.a}$$

and thus the isothermal parameter is given by $h = \frac{1}{2} \ln \frac{4}{(1+x^2+y^2)^2} = \ln(2) - \ln(1 + x^2 + y^2)$. We use Lemma 3.21 to see that this is a model for spherical geometry by computing:

$$h_{/11} = -\partial_x\{2x(1 + x^2 + y^2)^{-1}\} = -\{2(1 + x^2 + y^2) - 4x^2\}(1 + x^2 + y^2)^{-2},$$
$$h_{/22} = -\partial_y\{2y(1 + x^2 + y^2)^{-1}\} = -\{2(1 + x^2 + y^2) - 4y^2\}(1 + x^2 + y^2)^{-2},$$
$$K = -e^{-2h}(h_{/11} + h_{/22}) = \frac{(1+x^2+y^2)^2}{4}\frac{4+4x^2+4y^2-4x^2-4y^2}{(1+x^2+y^2)^2} = 1.$$

3.6.6 THE PSEUDO-SPHERE MODEL FOR HYPERBOLIC SPACE.

. Nikolai Lobachevsky was a Russian differential geometer who initiated the study of hyperbolic geometry; this field is often referred to as *Lobachevskian geometry*.

Nikolai Ivanovich Lobachevsky (1792–1856)

One model of hyperbolic space is given by the pseudo-sphere

$$S := \{x \in \mathbb{R}^3 : \langle x, x \rangle = -1\} \quad \text{where} \quad \langle x, y \rangle = x^1 y^1 + x^2 y^2 - x^3 y^3.$$

The orthogonal group $O(3, 1)$ acts transitively on S so S has constant Gaussian curvature. Let $T(x, y) = (x, y, \pm(1 + x^2 + y^2)^{\frac{1}{2}})$ parametrize the two components of the pseudo-sphere as graphs. We compute:

$$\partial_x T = (1, 0, \pm x(1 + x^2 + y^2)^{-\frac{1}{2}}), \quad \partial_y T = (0, 1, \pm y(1 + x^2 + y^2)^{-\frac{1}{2}}),$$
$$g_{11} = 1 - x^2(1 + x^2 + y^2)^{-1}, \qquad g_{12} = -xy(1 + x^2 + y^2)^{-1},$$
$$g_{22} = 1 - y^2(1 + x^2 + y^2)^{-1}.$$

Since $g_{ij} = \delta_{ij} + O(|x|^2)$, we use Lemma 3.13 to compute that the pseudo-sphere has constant Gaussian curvature -1; $R_{1221}(0, 0) = \frac{1}{2}\left(2g_{12/12} - g_{22/11} + g_{11/22}\right)(0, 0) = -1$. The isometry group is the component of $O(1, 2)$ which preserves the upper hyperboloid.

The hyperbolic plane

3.6.7 THE UNIT DISK MODEL FOR HYPERBOLIC SPACE. We now consider the analogue of stereographic projection. The line between $(x, y, 0)$ and $(0, 0, -1)$ is parametrized in the form $T(x, y) = t(x, y, 0) + (1 - t)(0, 0, -1)$. To ensure T takes values in the hyperboloid, we need that $\|T\|^2 = t^2(x^2 + y^2) - (1 - t)^2 = -1$. This shows that $t^2(x^2 + y^2 - 1) + 2t = 0$ so since we need $t \neq 0$, we have $t = 2(1 - x^2 - y^2)^{-1}$ and

$$T(x, y) = (2x, 2y, 1 + x^2 + y^2)/(1 - x^2 - y^2).$$

The hyperbolic stereographic projection

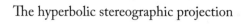

Let $\mathcal{E} = (1 - x^2 - y^2)^2$. We compute in a similar fashion as we did with the sphere that:

$$\partial_x T = (2 - 2y^2 + 2x^2, 4xy, 4x)/\mathcal{E},$$
$$\partial_y T = (4xy, 2 - 2x^2 + 2y^2, 4y)/\mathcal{E},$$
$$g_{11} = 4(1 - x^2 - y^2)^2/(1 - x^2 - y^2)^4, \qquad g_{12} = 0,$$
$$g_{22} = 4(1 - x^2 - y^2)^2/(1 - x^2 - y^2)^4.$$

This yields the unit disk model for hyperbolic space; we change the sign in Equation (3.6.a):

$$ds^2 := 4(dx^2 + dy^2)(1 - x^2 - y^2)^{-2}.$$

These are isothermal coordinates with $h = \ln(2) - \ln(1 - x^2 - y^2)$. By Lemma 3.21, the Gaussian curvature is given by $K = -e^{-2h}(h_{/11} + h_{/22})$. We compute:

$$
\begin{aligned}
h_{/11} &= \partial_x \{2x(1 - x^2 - y^2)^{-1}\} = \{2(1 - x^2 - y^2) + 4x^2\}(1 - x^2 - y^2)^{-2}, \\
h_{/22} &= \partial_y \{2y(1 - x^2 - y^2)^{-1}\} = \{2(1 - x^2 - y^2) + 4y^2\}(1 - x^2 - y^2)^{-2}, \\
K &= -\tfrac{1}{4}(1 - x^2 - y^2)^2(4 - 4x^2 - 4y^2 + 4x^2 + 4y^2)(1 - x^2 - y^2)^{-2} = -1.
\end{aligned}
$$

The geodesics in this model are the straight lines through the origin and the circles which are perpendicular to the boundary circle.

3.6.8 THE UPPER HALF PLANE MODEL FOR HYPERBOLIC SPACE. We give the upper half plane $M := \{(x, y) : y > 0\}$ the metric $ds^2 = y^{-2}(dx^2 + dy^2)$. Let $z = i\frac{w+1}{1-w}$; we solve the equation $z(1 - w) = iw + i$ for z to see the inverse map is given by $w = \frac{z-i}{z+i}$. We compute:

$$
dz = 2i(1 - w)^{-2}dw, \quad d\bar{z} = -2i(1 - \bar{w})^{-2}d\bar{w}, \quad dz \circ d\bar{z} = 4|1 - w|^{-4}dw \circ d\bar{w},
$$

$$
y = \Im\left\{i\frac{w + 1}{-w + 1}\right\} = |1 - w|^{-2}\Re((w + 1)(-\bar{w} + 1)) = |1 - w|^{-2}(1 - |w|^2),
$$

$$
y^{-2}dz \circ d\bar{z} = 4(1 - |w|^2)^{-2}dw \circ d\bar{w}.
$$

Thus $|w| < 1 \leftrightarrow y > 0$, $|w| > 1 \leftrightarrow y < 0$, and the circle $|w| = 1$ goes to $\mathbb{R} \cup \infty$. Furthermore the Jacobian is non-singular for $w \neq 1$. Thus this provides an isometry of the interior of the unit disk with the metric $4(1 - |w|^2)^{-2}dw \circ d\bar{w}$ to the upper half plane with the metric $y^{-2}(dz \circ d\bar{z})$. The isothermal parameter for the upper half plane model is given by setting $h = \frac{1}{2}\ln(y^{-2}) = -\ln(y)$ so

$$
h_{/11} = 0, \quad h_{/22} = -\partial_y(y^{-1}) = y^{-2}, \quad K = -e^{-2h}(h_{/11} + h_{/22}) = -1.
$$

The geodesics in this geometry are straight lines perpendicular to the x-axis and circles perpendicular to the x-axis. The *special linear group* $\mathrm{SL}(2, \mathbb{R})$ is the group of 2×2 real matrices of determinant 1. Set:

$$
T_A z := \frac{az + b}{cz + d} \quad \text{for} \quad A = \begin{pmatrix} a & b \\ c & d \end{pmatrix} \in \mathrm{SL}(2, \mathbb{R}).
$$

The map $z \to Az$ is an isometry of this geometry and every orientation preserving isometry arises in this fashion. The full group of isometries is then generated by adding the isometry $z \to -\bar{z}$.

3.6.9 THE TRACTRIX MODEL FOR HYPERBOLIC SPACE. There is a partial realization called *tractrix* of the hyperbolic plane in $\mathbb{R}^3$ with the Euclidean inner product as a surface of revolution that is singular along the line $u = 0$ (Beltrami 1868). It is given by

$$(x = \text{sech}(u)\cos(v), y = \text{sech}(u)\sin(v), z = u - \tanh(u))$$

We have $g_{uu} = \tanh^2(u)$, $g_{uv} = 0$, $g_{vv} = \text{sech}^2(u)$ while $\text{II}_{uu} = -\text{sech}(u)\tanh(u)$, $\text{II}_{uv} = 0$, and $\text{II}_{vv} = \text{sech}(u)\tanh(u)$. Thus $K = -1$.

3.6.10 THE LORENTZ SPHERE. Let $\langle x, y \rangle = x^1 y^1 + x^2 y^2 - x^3 y^3$ be the Lorentzian inner product on $\mathbb{R}^3$. We obtained a model for the hyperbolic plane by considering the pseudo-sphere $\langle x, x \rangle = -1$; this had signature $(0, 2)$. We now consider the *Lorentz sphere* $S = \{x : \langle x, x \rangle = 1\}$; this has signature $(1, 1)$ and is a hyperboloid of one sheet which may be parametrized by setting $T(u, v) := (\cosh(u)\cos(v), \cosh(u)\sin(v), \sinh(u))$. We compute:

$$\partial_u T = (\sinh(u)\cos(v), \sinh(u)\sin(v), \cosh(u)),$$
$$\partial_v T = (-\cosh(u)\sin(v), \cosh(u)\cos(v), 0),$$
$$v = (\cosh(u)\cos(v), \cosh(u)\sin(v), \sinh(u)),$$
$$g_{uu} = -1, \quad g_{uv} = 0, \quad g_{vv} = \cosh^2 u,$$
$$\text{II}_{uu} = 1, \quad \text{II}_{uv} = 0.$$

Consequently, $K = \det(\text{II})/\det(g) = +1$.

The Lorentz sphere

3.6.11 THE GEODESIC CURVATURE. Let $\gamma(t)$ be a curve parametrized by arc-length so $\|\dot{\gamma}\| = 1$. Let $v(t)$ be a unit normal to the curve. The *geodesic curvature* is defined to be

$$\kappa_g(\gamma) := g(\nabla_{\dot{\gamma}}\dot{\gamma}, v).$$

Since γ has constant length, $0 = \partial_t g(\dot{\gamma}, \dot{\gamma}) = 2g(\nabla_{\dot{\gamma}}\dot{\gamma}, \dot{\gamma})$ and thus $\ddot{\gamma}$ is perpendicular to $\dot{\gamma}$. This implies $\ddot{\gamma} = \kappa_g(\gamma) \cdot v$ and thus γ is a geodesic if and only if $\kappa_g(\gamma) = 0$.

Let $\gamma(t) = (r\cos(t/r), r\sin(t/r))$ parametrize the circle of radius r about the origin. Then $\dot{\gamma}(t) = (-\sin(t/r), \cos(t/r))$ is a unit vector so γ is parametrized by arc length. The (inward) unit normal to the circle is given by $v(t) = (-\cos(t/r), -\sin(t/r))$ and the geodetic curvature is given by

$$\kappa_g = \ddot{\gamma}(t) \cdot v(t) = \tfrac{1}{r}(-\cos(t/r), -\sin(t/r)) \cdot (-\cos(t/r), -\sin(t/r)) = \tfrac{1}{r}.$$

If γ is an arbitrary curve in the plane, and if C_r is the best circle approximating γ, then $|\kappa_g| = \pm\tfrac{1}{r}$ where the $\pm$ sign reflects the choice of normal.

3.6.12 ANGLE SUM FORMULAS. A unit speed curve in the plane is a geodesic if and only if $\ddot{\gamma} = 0$, i.e., γ is a straight line. If T is a geodesic triangle with angles θ_i at the vertices, then one has from Euclid that:

$$\theta_1 + \theta_2 + \theta_3 = \pi. \tag{3.6.b}$$

Let S^2 be the unit sphere in $\mathbb{R}^3$ with the induced metric. Let P be a plane through the origin and let $\gamma = P \cap S^2$ be a great circle in S^2. Reflection in P defines an isometry of $\mathbb{R}^3$ whose fixed point set is P; Lemma 3.10 then shows γ is a geodesic. Thus the great circles comprise the geodesics of S^2. In 1603 Thomas Harriot showed that if T is a triangle in the unit sphere with angles $\{\theta_1, \theta_2, \theta_3\}$, then

$$\text{Area}(T) = \theta_1 + \theta_2 + \theta_3 - \pi. \tag{3.6.c}$$

His result is unpublished; the first published result seems to be due to Albert Girard. In other words, the sum of the angles of a spherical triangle exceeds π and the excess measures the area.

In hyperbolic geometry (i.e., on a surface of constant sectional curvature -1), the sign in Equation (3.6.c) is reversed. If T is a triangle whose edges are geodesics in the hyperbolic plane, then Lampert [30] showed in 1766 that

$$\text{Area}(T) = \pi - \theta_1 - \theta_2 - \theta_3. \tag{3.6.d}$$

3.6.13 THE GAUSS–BONNET THEOREM. Equations (3.6.b), (3.6.c), and (3.6.d) are consequences of the *Gauss–Bonnet Theorem* which is named after Carl Friedrich Gauss (who first investigated the Gaussian curvature K) and Pierre Ossian Bonnet (who published a special case of the Gauss–Bonnet Theorem in 1848); the general case seems to be due to von Dyck (1888) who discussed the global version:

Theorem 3.23   *Let (M, g) be a 2-dimensional Riemannian manifold with piecewise smooth boundary ∂M. Let $K = R_{1221}$ be the Gaussian curvature, let κ_g be the geodesic curvature, and let α_i be the interior angles at which consecutive smooth boundary components meet. Then*

$$2\pi \cdot \chi(M) = \int_M K + \int_{\partial M} \kappa_g + \sum_i (\pi - \alpha_i).$$

This result is due to the German mathematician Johann Carl Friedrich Gauss (1777–1855), to the French mathematician Pierre Ossian Bonnet (1819–1892), and to Walther Franz Anton von Dyck (1856–1934).

Gauss   Bonnet

If we specialize Ω to a triangle in the plane, then $K = 0$ and $\kappa_g = 0$ and we have

$$\int_\Omega 0\,dA + \sum_{i=1}^{3}(\pi - \theta_i) = 2\pi, \quad \text{or equivalently,} \quad \theta_1 + \theta_2 + \theta_3 - \pi = 0$$

and we recover the result of Euclid given in Equation (3.6.b). If we specialize Ω to a spherical triangle with angles θ_i, then $K = +1$ and $\kappa_g = 0$ and we have

$$\int_\Omega 1\,dA + \sum_{i=1}^{3}(\pi - \theta_i) = 2\pi, \quad \text{or equivalently,} \quad \text{Area}(\Omega) = \theta_1 + \theta_2 + \theta_3 - \pi,$$

so we recover the result of Harriott given in Equation (3.6.c). On the other hand if Ω is a hyperbolic triangle, then $K = -1$, then we may establish Equation (3.6.d):

$$\int_\Omega (-1)\,dA + \sum_{i=1}^{3}(\pi - \theta_i) = 2\pi, \quad \text{or equivalently,} \quad -\text{Area}(\Omega) = \theta_1 + \theta_2 + \theta_3 - \pi.$$

Proof. We first examine the local theory and take $M = \mathbb{R}^2$. We have global coordinates (x, y) on M which we use to fix the orientation. Let Ω be a subdomain of M with piecewise smooth boundary. Let $\{e_1, e_2\}$ be an orthonormal frame. Expand

$$\nabla e_1 = \omega e_2 \quad \text{and} \quad \nabla e_2 = -\omega e_1 \quad \text{for} \quad \omega = \omega_x dx + \omega_y dy.$$

We take the usual orientation $dx \wedge dy$ for $\mathbb{R}^2$ and assume $e^1 \wedge e^2$ is the orientation as well. Recall that $g = \det(g_{ij})^{\frac{1}{2}}$ so the Riemannian measure is $g\,dx\,dy$. The Gaussian curvature is then characterized by the equation

$$
\begin{aligned}
K|\,dvol\,| &= g^{-1}g((\nabla_x\nabla_y - \nabla_y\nabla_x)e_2, e_1) \cdot g\,dx \wedge dy \\
&= (-\partial_x\omega_y + \partial_y\omega_x)dx \wedge dy = -d\omega.
\end{aligned}
$$

Assume that $\partial\Omega$ is smooth. Use arc length to parametrize the boundary, oriented counterclockwise, by $\gamma(t)$. We may express $\dot\gamma$ and the normal ν in the form:

$$\dot\gamma(t) = -\sin(\theta(t))e_1 + \cos(\theta(t))e_2 \quad \text{and} \quad \nu(t) = -\cos(\theta(t))e_1 - \sin(\theta(t))e_2.$$

(This is modeled on taking the circle $\gamma(t) = (\cos(t), \sin(t))$ and using the inward unit normal in flat space). We compute

$$\nabla_{\dot{\gamma}}\dot{\gamma} = \{-\cos(\theta(t))e_1 - \sin(\theta(t))e_2\}\dot{\theta}(t) + \{-\cos(\theta(t))e_1 - \sin(\theta(t))e_2\}\omega(\dot{\gamma}),$$
$$\kappa_g = g(\nabla_{\dot{\gamma}}\dot{\gamma}, v) = \dot{\theta} + \omega(\dot{\gamma}).$$

Thus $\int_{\partial\Omega} \kappa_g\, dt = \int_{\partial\Omega} \dot{\theta}\, dt + \int_{\partial\Omega} \omega$ so we may apply Stokes' Theorem (which holds even if the boundary has corners) to see:

$$\int_{\Omega} K|\,\mathrm{dvol}\,| + \int_{\partial\Omega} \kappa_g\, ds = -\int_{\Omega} d\omega + \int_{\partial\Omega} \omega + \int_{\partial\Omega} \dot{\theta}\, dt = \int_{\partial\Omega} \dot{\theta}\, dt = 2\pi\,.$$

Now if the boundary has corners, then the angular parameter θ has jumps at the corners. If P_i is such a corner and if the exterior angle is $\beta_i = \pi - \alpha_i$, we have

$$\int_{\partial\Omega} \dot{\theta}\, dt + \sum_i (\pi - \alpha_i) = 2\pi$$

is a full turn. This gives rise to the desired formula

$$\int_{\Omega} K|\,\mathrm{dvol}\,| + \int_{\partial\Omega} \kappa_g\, ds + \sum_i (\pi - \alpha_i) = 2\pi\,,$$

since $\chi(\Omega) = 1$. The fact that Ω was geodesically convex plays a role only in seeing that the total turn is 2π; it works equally well if Ω is compact and if $\partial\Omega$ has only one component. If Ω has holes, then the outer component gives rise to 2π and each inner component gives rise to -2π. Since the Euler characteristic is 1 minus the number of inner boundary components, the desired result follows in this special case. The general case now follows by a cutting and pasting argument; we omit the details in the interests of brevity. □

3.7   THE CHERN–GAUSS–BONNET THEOREM

Theorem 3.23 has been generalized to the higher-dimensional setting by Chern [10]. For the moment, we shall assume that (M, g) is a smooth compact Riemannian manifold without boundary; we shall discuss the boundary correction term presently. Let m be even and let $\{e_1, \ldots, e_m\}$ be a frame for (M, g). Let 2ℓ be even and set

$$E_m := \frac{1}{(8\pi)^\ell \ell!} R_{i_1 i_2 j_2 j_1} \cdot \cdots \cdot R_{i_{\ell-1} i_\ell j_\ell j_{\ell-1}} g(e^{i_1} \wedge \cdots \wedge e^{i_\ell}, e^{j_1} \wedge \cdots \wedge e^{j_\ell})\,. \qquad (3.7.a)$$

Set $|\,\mathrm{dvol}\,| = \det(g_{\alpha,ij})^{\frac{1}{2}} dx^1 \cdot \cdots \cdot dx^m$. We refer to Chern [10] for the proof of the following result. We also refer to contemporaneous related work by Allendoerfer and Weil [2].

S. Chern (1911–2004)

Many proofs have been given subsequently of this result.

Theorem 3.24 *Let (M, g) be a compact Riemannian manifold without boundary of dimension m. If m is odd, the Euler–Poincaré characteristic $\chi(M)$ vanishes. If m is even,*

$$\chi(M) = \int_M E_m |\,\mathrm{dvol}\,| .$$

3.7.1 THE CHERN–GAUSS–BONNET THEOREM IN INDEFINITE SIGNATURE.
Chern [11] and Avez [5] subsequently extended Theorem 3.24 to the indefinite setting. In the pseudo-Riemannian setting, the volume element is given by

$$|\,\mathrm{dvol}\,| = |\det(g_{ij})|^{\frac{1}{2}} dx^1 \wedge \cdots \wedge dx^m .$$

This is a measure and not a differential form; no assumption of orientability is made.

Theorem 3.25 *Let (M, g) be a compact pseudo-Riemannian manifold of signature (p, q) without boundary of even dimension m. If p is odd, then $\chi(M)$ vanishes. If p is even, then*

$$\chi(M) = (-1)^{p/2} \int_M E_m |\,\mathrm{dvol}\,| .$$

Proof. We follow the discussion in Gilkey and Park [18]. Let $g \in S^2(T^*M) \otimes \mathbb{C}$ be a complex-valued symmetric bilinear form on the tangent space. We assume $\det(g) \neq 0$ as a non-degeneracy condition. The Levi–Civita connection and curvature tensor may then be defined. We set $\mathrm{dvol}(g) := \det(g_{ij})^{\frac{1}{2}} dx^1 \wedge \cdots \wedge dx^m$; we do not take the absolute value. There is a subtlety here since, of course, there are two branches of the square root function. We shall ignore this for the moment in the interests of simplifying the argument and return to this point in a moment. We use Equation (3.7.a) to define the Euler form and consider $\int_M E_m(g)\,\mathrm{dvol}(g)$. Let $h \in S^2(T^*M) \otimes \mathbb{C}$ give a perturbation of the metric. Consider the 1-parameter family $g_\epsilon := g + \epsilon h$ and differentiate with respect to the parameter ϵ. We then integrate by parts to define the Euler–Lagrange equations:

$$\partial_\epsilon \left\{ \int_M E_m(g + \epsilon h)\,\mathrm{dvol}(g + \epsilon h) \right\} \bigg|_{\epsilon=0} = \int_M g((TE_m)(g), h)\,\mathrm{dvol}(g)$$

where $TE_m(g) \in S^2(T^*M) \otimes \mathbb{C}$ is the *transgression* of the Pfaffian E_m. This is defined in any dimension. Although a priori TE_m can involve the 4th derivatives of the metric, Berger [7] conjectured it only involved curvature; this was subsequently verified by Kuz'mina [26] and Labbi [27–29]. We also refer to related work in Gilkey, Park and Sekigawa [19]. What is important for us, however, is that Theorem 3.24 shows that $TE_m(g) = 0$ if $\dim(M) = m$ and if g is Riemannian, i.e., if g and h are real since the integrated invariant $\int_M E_m(g + \epsilon h)\, \text{dvol}(g + \epsilon h) = \chi(M)$ is independent of (ϵ, h). The invariant TE_m is locally computable. In any coordinate system, it is polynomial in the derivatives of the metric, the components of the metric, and $\det(g_{ij})^{-\frac{1}{2}}$. Thus it is holomorphic in the metric. Thus vanishing identically when g is real implies it vanishes identically when g is complex. Consequently, the Euler–Lagrange equations vanish. In particular, if g_ϵ is a smooth 1-parameter family of complex metrics starting with a real metric, then we can define a branch of the square root function along this family so that the following integral is independent of ϵ:

$$\int_M E_m(g_\epsilon)\, \text{dvol}(g_\epsilon).$$

We apply Lemma 3.2 to find an orthogonal direct sum decomposition $TM = V_- \oplus V_+$ where V_- is timelike with $\dim(V_-) = p$ and where V_+ is spacelike with $\dim(V_+) = q$. This decomposes $g = g_- \oplus g_+$ where g_- is negative definite on V_- and g_+ is positive definite on V_+. For $\epsilon \in [0, \pi]$, define:

$$g_\epsilon = -e^{\epsilon\sqrt{-1}} g_- \oplus g_+ \,.$$

Then $g_0 = -g_- \oplus g_+$ is a Riemannian metric while $g_\pi = g_- \oplus g_+$ is the given pseudo-Riemannian metric. We have

$$\det(g_\epsilon) = e^{\epsilon\sqrt{-1}p} \det(-g_-)\det(g_+) = e^{\epsilon\sqrt{-1}p} \det(g_0)$$

so the branch of the square root function along the deformation from g_0 to g_π is given by $e^{\epsilon\sqrt{-1}p/2}$. Thus if p is odd, this is purely imaginary for g_π and hence the Euler characteristic vanishes. If p is even, then $\det(g_0) = \det(g_\pi)$ so $(-1)^{p/2}\det(g_0)^{\frac{1}{2}} = (-1)^{p/2}\det(g_\pi)^{\frac{1}{2}}$. This completes the proof. □

3.7.2 EXAMPLES. Let (M, g_0) be a Riemann surface. Let $g = -g_0$ have signature $(2, 0)$. Then

$$^g\nabla = {}^{g_0}\nabla \quad \text{so} \quad {}^gR_{ijk}{}^\ell = {}^{g_0}R_{ijk}{}^\ell \quad \text{and} \quad \tau(g) = g^{jk}\, {}^gR_{ijk}{}^i = -g_0^{jk}\, {}^{g_0}R_{ijk}{}^i = -\tau(g_0)\,.$$

As $|\,\text{dvol}\,|(g_0) = |\,\text{dvol}\,|(g)$, one must change the sign in the Gauss–Bonnet Theorem:

$$\chi(M) = -\frac{1}{4\pi}\int_M \tau(g)|\,\text{dvol}\,|(g)\,.$$

In dimension four, if $(M, g) = (M_1, g_1) \times (M_2, g_2)$, then the Gauss–Bonnet Theorem decouples and we have $\chi(M) = \chi(M_1)\chi(M_2)$ and $E_4(g) = E_2(g_1)E_2(g_2)$. Thus we will not need to change the sign in signature $(4, 0)$ or $(0, 4)$ but we will need to change the sign in signature $(2, 2)$. The fact that the Euler characteristic vanishes if p and q are both odd is not, of course, new but follows from standard characteristic class theory.

It is possible to generalize the Gauss–Bonnet Theorem to manifolds with boundary; in the Riemannian setting, this is due to Chern [10] and in the pseudo-Riemannian setting to Alty [3]. We introduce a bit of additional notation. Let (M, g) be a pseudo-Riemannian manifold of signature (p, q) with smooth boundary ∂M. We assume the restriction of g to the boundary is non-degenerate and let v be the inward pointing unit normal. Let $\{v, e_2, \dots, e_m\}$ be a local orthonormal frame field. Let $L_{ab} := g(\nabla_{e_a} e_b, v)$ be the second fundamental form. We sum over indices a_i and b_i ranging from 2 to m to define:

$$F_{m,v} := \left\{ \frac{R_{a_1 a_2 b_2 b_1} \cdots \cdot R_{a_{2v-1} a_{2v} b_{2v} b_{2v-1}} L_{a_{2v+1} b_{2v+1}} \cdots \cdot L_{a_{m-1} b_{m-1}}}{(8\pi)^v v! \operatorname{Vol}(S^{m-1-2v})(m-1-2v)!} \right\}$$
$$\times g(e^{a_1} \wedge \cdots \wedge e^{a_{m-1}}, e^{b_1} \wedge \cdots \wedge e^{b_{m-1}}).$$

Note that if m is odd, then $\chi(M) = \frac{1}{2}\chi(\partial M)$ so we may apply Theorem 3.25 to compute $\chi(\partial M)$ and thereby $\chi(M)$ in terms of curvature. We therefore assume m is even.

Theorem 3.26   *Let (M, g) be a compact smooth manifold pseudo-Riemannian manifold of even dimension m and signature (p, q) which has smooth boundary ∂M. If p is odd, then $\chi(M) = 0$. Otherwise*

$$\chi(M) = (-1)^{p/2} \left\{ \int_M E_m(g) |\operatorname{dvol}|(g_M) + \sum_v \int_{\partial M} F_{m,v} |\operatorname{dvol}|(g_{\partial M}) \right\}.$$

Proof. If (M, g) is Riemannian, this is well-known (see, for example, the discussion in Gilkey [17] using heat equation methods and extending the work of Patodi [35] to the case of manifolds with boundary). So the trick is to extend it to the pseudo-Riemannian setting. Again, we will use analytic continuation. But there is an important difference. We fix a non-zero vector field X which is inward pointing on the boundary. We consider a smooth 1-parameter of complex variations g_ϵ so that $g_\epsilon(X, X) \neq 0$ and so that $X \perp T(\partial M)$ with respect to g_ϵ. We may then set $v = X \cdot g_\epsilon(X, X)^{-\frac{1}{2}}$. In the expression for $F_{m,v}$, there are an odd number of terms which contain L and hence $g_\epsilon(X, X)^{-\frac{1}{2}}$. Given a pseudo-Riemannian metric g, we construct g_0 as above. Let $g_{ab}^{\partial M}$ and $g_{0,ab}^{\partial M}$ be the restriction of the metrics g and g_0, respectively, to the boundary. If X is spacelike, then

$$\det(g_{ab}^{\partial M}) = (-1)^p \det(g_{0,ab}^{\partial M})$$

and the analysis proceeds as previously; the whole subtlety arising from the square root of the determinant. If X is timelike, then

$$\det(g_{ab}^{\partial M}) = (-1)^{p-1} \det(g_{0,ab}^{\partial M}).$$

However $g(X, X) = -g_0(X, X)$ and thus once again, we must take the square root of $(-1)^p$ in the analytic continuation. Apart from this, the remainder of the argument is the same as that used to prove Theorem 3.25. □

Bibliography

[1] J. F. Adams, "Vector fields on spheres," *Bull. Amer. Math. Soc.* **68** (1962), 39–41. DOI: 10.1090/S0002-9904-1962-10693-4.

[2] C. Allendoerfer and A. Weil, "The Gauss-Bonnet theorem for Riemannian polyhedra," *Trans. Am. Math. Soc.* **53** (1943), 101–129. DOI: 10.1090/S0002-9947-1943-0007627-9. 126

[3] L. J. Alty, "The generalized Gauss-Bonnet-Chern theorem," *J. Math. Phys.* **36** (1995), 3094–3105. DOI: 10.1063/1.531015. 129

[4] M. F. Atiyah, "K-theory. Lecture notes by D. W. Anderson," W. A. Benjamin, Inc., New York-Amsterdam (1967). 59

[5] A. Avez, "Formule de Gauss-Bonnet-Chern en métrique de signature quelconque," *C. R. Acad. Sci. Paris* **255** (1962), 2049–2051. 127

[6] J. Baum, "Elements of point set topology," Prentice Hall, Englewood Cliffs, N. J. (1964). 1

[7] M. Berger, "Quelques formulas de variation pour une structure riemanniene," *Ann. Sci. Éc. Norm. Supér.* **3** (1970), 285–294. 128

[8] A. L. Besse, "Einstein manifolds," *Ergebnisse der Mathematik und ihrer Grenzgebiete. 3. Folge* **10**. Springer-Verlag, Berlin (1987). 95

[9] M. P. do Carmo, "Geometria riemanniana," *Projeto Euclides* **10**. Instituto de Matematica Pura e Aplicada, Rio de Janeiro (1979). 110

[10] S. Chern, "A simple proof of the Gauss-Bonnet formula for closed Riemannian manifolds," *Ann. Math.* **45** (1944), 747–752. DOI: 10.2307/1969302. 126, 129

[11] S. Chern, "Pseudo-Riemannian geometry and the Gauss-Bonnet formula," *An. Acad. Brasil. Ci.* **35** (1963), 17–26. 127

[12] A. Clairaut, "Théorie de la figure de la terre, tirée des principes de l'hydrostatique," Paris (1743). 36

[13] G. Cramer, "Introduction à l'Analyse des lignes Courbes algébriques," *Genève: Frères Cramer & Cl. Philbert* (1750) (see pages 656–659). 6

[14] S. Dickson, "Klein Bottle Graphic" (1991) 57
http://library.wolfram.com/infocenter/MathSource/4560/.

[15] L. Eisenhart, "Riemannian Geometry," Princeton University Press, Princeton, N.J. (1949).

[16] G. Frobenius, "Über das Pfaffsche probleme," *J. für Reine und Agnew. Math.* **82** (1877), 230–315. 65

[17] P. Gilkey, "Invariance theory, the heat equation, and the Atiyah-Singer index theorem," *Studies in Advanced Mathematics*, CRC Press, Boca Raton (1995). 129

[18] P. Gilkey and J.H. Park, "Analytic continuation, the Chern-Gauss-Bonnet theorem, and the Euler-Lagrange equations in Lovelock theory for indefinite signature metrics", *Journal of Geometry and Physics* (2015), pp. 88–93. DOI: 10.1016/j.geomphys.2014.11.006. 127

[19] P. Gilkey, J. H. Park, and K. Sekigawa, "Universal curvature identities," *Differ. Geom. Appl.* **29** (2011), 770–778. DOI: 10.1016/j.difgeo.2011.08.005. 128

[20] A. Gray, "The volume of a small geodesic ball of a Riemannian manifold," *Michigan Math. J.* **20** (1974), 329–344. DOI: 10.1307/mmj/1029001150.

[21] H. Hopf and W. Rinow, "Über den Begriff der vollständigen differentialge-ometrischen Fläche," *CommentarII Mathematici Helvetici* **3** (1931), 209–225. DOI: 10.1007/BF01601813. 108

[22] J. Jost, "Riemannian geometry and geometric analysis," *Universitext*, Springer-Verlag, Berlin (2002). DOI: 10.1007/978-3-662-04745-3. 67

[23] M. Karoubi, "K-theory, an introduction," *Grundlehren der mathematischen Wissenschaften* **226**, Springer Verlag (Berlin) (1978). 59

[24] S. Kobayashi and K. Nomizu, "Foundations of Differential Geometry vol. I and II," *Wiley Classics Library*. A Wyley-Interscience Publication, John Wiley & Sons, Inc., New York (1996). 95

[25] O. Kowalski and M. Sekizawa, "Natural lifts in Riemannian geometry," *Variations, geometry and physics*, 189–207, Nova Sci. Publ., New York (2009). 69

[26] G. M. Kuz'mina, "Some generalizations of the Riemann spaces of Einstein," *Math. Notes* **16** (1974), 961–963; translation from *Mat. Zametki* **16** (1974), 619–622. DOI: 10.1007/BF01104264. 128

[27] M.-L. Labbi, "Double forms, curvature structures and the (p, q)-curvatures," *Trans. Am. Math. Soc.* **357** (2005), 3971–3992. DOI: 10.1090/S0002-9947-05-04001-8. 128

[28] M.-L. Labbi, "On Gauss-Bonnet Curvatures," SIGMA, *Symmetry Integrability Geom. Methods Appl.* **3**, Paper 118, 11 p., electronic only (2007). DOI: 10.3842/SIGMA.2007.118.

[29] M.-L. Labbi, "Variational properties of the Gauss-Bonnet curvatures," *Calc. Var. Partial Differ. Equ.* **32** (2008), 175–189. DOI: 10.1007/s00526-007-0135-4. 128

[30] J. Lambert, "Theorie der parallillinien," Leipzig (1766). 124

[31] T. Levi-Civita, "Nozione di parallelismo in una varietà qualunque e consequente specificazione geometrica della curvatura Riemanniana," *Rend. Circ. Mat. Palermo* **42** (1917), 73–205. DOI: 10.1007/BF03014898. 100

[32] J. Munkres, "Elementary differential topology," Lectures given at Massachusetts Institute of Technology, Fall, 1961. Annals of Mathematics Studies, No. 54 Princeton University Press, Princeton, N.J. (1963). 58

[33] S. B. Myers, "Riemannian manifolds with positive mean curvature," *Duke Mathematical Journal* **8** (1941), 401–404; DOI: 10.1215/S0012-7094-41-00832-3. 115

[34] B. O'Neill, "Semi-Riemannian geometry. With applications to relativity," *Pure and Applied Mathematics* **103**, Academic Press, Inc., New York (1983). 113

[35] V. K. Patodi, "Curvature and the eigenforms of the Laplace operator," *J. Differential Geometry* **5** (1971), 233–249. 129

[36] W. Rudin, "Principles of mathematical analysis," McGraw-Hill Book Co., New York-Auckland-Düsseldorf (1976). 1, 23

[37] E. Spanier, "Algebraic topology," Berlin: Springer-Verlag (1995). 55, 86

[38] M. Spivak, "Calculus on manifolds. A modern approach to classical theorems of advanced calculus," W. A. Benjamin, Inc., New York-Amsterdam (1965). 7, 12

[39] R. Stong, "Notes on cobordism theory". Mathematical notes Princeton University Press, Princeton, N.J.; University of Tokyo Press, Tokyo (1968). 78

[40] H. Whitney, "Differentiable manifolds," *Annals of Math.* **37** (1936), 645–680. DOI: 10.2307/1968482. 58

[41] J. Wolf, "Spaces of constant curvature," American Mathematical Society, Providence, RI (2011). 55, 113

[42] K. Yano and S. Ishihara, "Tangent and cotangent bundles," *Pure and Applied Mathematics* **16**, Marcel Dekker Inc., New York (1973). 67, 68

Authors' Biographies

PETER B GILKEY

Peter B Gilkey[1] is a Professor of Mathematics and a member of the Institute of Theoretical Science at the University of Oregon. He is a fellow of the American Mathematical Society and is a member of the editorial board of Results in Mathematics, J. Differential Geometry and Applications, and J. Geometric Analysis. He received his Ph.D. in 1972 from Harvard University under the direction of of L. Nirenberg. His research specialties are Differential Geometry, Elliptic Partial Dif-

ferential Equations, and Algebraic topology. He has published more than 250 research articles and books.

JEONGHYEONG PARK

JeongHyeong Park[2] is a Professor of Mathematics at Sungkyunkwan University and is an associate member of the KIAS (Korea). She received her Ph.D. in 1990 from Kanazawa University in Japan under the direction of H. Kitahara. Her research specialties are spectral geome-try of Riemannian submersion and geometric structures on manifolds like eta-Einstein manifolds and H-contact manifolds. She organized the geometry section of AMC 2013 (The Asian Mathematical

Conference 2013) and the ICM 2014 satellite conference on Geometric analysis. She has published more than 71 re-search articles and books.

[1]Mathematics Department, University of Oregon, Eugene OR 97403 U.S.
  email: gilkey@uoregon.edu
[2]Mathematics Department, Sungkyunkwan University, Suwon, 440-746, Korea
  email: parkj@skku.edu

RAMÓN VÁZQUEZ-LORENZO

Ramón Vázquez-Lorenzo[3] is a member of the research group in Riemannian Geometry at the Department of Geometry and Topology of the University of Santiago de Compostela (Spain). He is a member of the Spanish Research Network on Relativity and Gravitation. He received his Ph.D. in 1997 from the University of Santiago de Compostela under the direction of E. García-Río. His research focuses mainly on Differential Geometry with special emphasis on the study of the curvature and the algebraic properties of curvature operators in the Lorentzian and in the higher signature settings. He has published more than 50 research articles and books.

[3]Department of Geometry and Topology, Faculty of Mathematics, University of Santiago de Compostela, 15782 Santiago de Compostela, Spain.
email: ravazlor@edu.xunta.es

Index